REWILD
YOUR
GARDEN

FRANCES TOPHILL

REWILD YOUR GARDEN

Create a Haven for Birds,
Bees and Butterflies

greenfinch

First published in Great Britain in 2020 by

Greenfinch
An imprint of Quercus Editions Ltd
Carmelite House
50 Victoria Embankment
London EC4Y 0DZ

An Hachette UK company

A CIP catalogue record for this book is available from the British Library

HB ISBN 978-1-52941-025-9
E-BOOK ISBN 978-1-52941-024-2

10 9 8 7 6 5

Design by Tokiko Morishima
Cover and interior artwork by Jo Parry

Printed and bound in Italy by L.E.G.O SpA

Papers used by Greenfinch are from well-managed forests and other responsible sources.

CONTENTS

INTRODUCTION

As a child I remember being astounded, disbelieving even, when I learned that human beings are in fact animals. Just like the hares, swallows, dragonflies and other species we know so well, we evolved as a part of the natural world, deeply and inextricably connected to it. Somewhere along the line though, we have removed ourselves from nature. Our cycles are not the natural cycles and over thousands of years our way of life has become more and more separate from the world around us. That, of course, was the reason for my surprise; because we live so differently, almost unrecognizably so, from the rest of the animal kingdom that we are a part of.

The main difference between us and other animals is our desire and ability to control our environment, both in our homes and in the wider landscape. Over thousands of years we have removed forests to make better land for our livestock and crops, eliminating predators in the process, and smoothed huge swathes of countryside simply to make them more pleasant and easier to walk across. We've drained wetlands to create farmlands and used the natural resources we've found in the ground irrespective of the damage it has done to the local ecosystem. Historically, in our own little patches of land, there has been huge pressure on people to keep our gardens 'under control'. We must win the battle with the weeds, make sure the lawn contains only grass (and even then only approved species of grass) and

arrange our plants in groupings of threes and fives because they look more pleasing. But what has all this controlling, both large- and small-scale, done for the rest of the species that share the world with us? The answer is a dramatic increase in the rise of threatened species, a reduction in biodiversity and a planet that is facing an increasingly, and at times alarmingly, unpredictable climate. Wildlife needs as much help as we can give it.

In recent years rewilding has been seen as a possible solution, and become a bit of a buzzword. There have been trials the world over into whether it is possible to reverse man's influence and allow the natural landscape and all its species to recover. There is also huge debate about the level of intervention that we should make, whether none at all, which is rewilding in its purest form, or some small nudges in the right direction to help nature on her way.

Could and should we apply some of these principles in our own gardens? Completely rewilding a garden poses some problems. Our gardens are generally only ours for a few years, while true rewildling takes hundreds, if not thousands, of years. And, when seen in comparison to huge areas of woodland or wetland for example, gardens are extremely small and often surrounded by urban development or other gardens with limited plant species. The usual mechanisms of seed dispersal that act on the wider landscape, such as water courses moving through, various birds and wildlife carrying in and spreading seed, and wind blowing in seeds and spores, are not so easy in built-up areas. As a result you would expect to see brambles galore and a host of sycamore saplings – not exactly a diverse ecosystem for wildlife to inhabit, and definitely not an attractive garden!

In our gardens we often talk about wildlife gardening, which supports our beloved and threatened native species such as hedgehogs, barn owls, voles, bats and in recent years, rabbits and hares, as well as orchids and other endangered plants. The key difference between wildlife gardening and rewilding is that in the former, the gardener selects specific 'good' birds, animals, plants and insects to foster – providing them with food sources and nesting spots, along with drinking, bathing and spawning sites. Rewilding in its purest form is about simply stopping intervening at all, so that a rich and diverse natural balance will be restored. The truth is that in a garden setting, a combination of wildlife gardening and rewilding (supporting here and relaxing some control there) is probably the most effective way of giving nature a helping hand, while not compromising too much on your own sense of aesthetic. After all, it is your space (for now, at least) and you want to enjoy it too!

I want to show every gardener some practical ways to encourage biodiversity, and in the process foster lots of species (not just birds, bees, butterflies and hedgehogs but also many, many others) while still having a beautiful space to sit in and enjoy. I am fundamentally a plants person, but I also have a deep love and respect for lichen, fungus and algae that live in such delicate harmony within our landscape and provide food and habitats to a host of creatures. These, along with the plants themselves, are often overlooked when we think about conservation. A huge range of insects beyond the much-loved honeybee are also crucial for supporting a rich and diverse level of wildlife – the woodlouse, the ant and the earwig are all essential in the garden food chain and support numerous other species. My goal is to give space for all of these unsung heroes who play a vital supporting role within our garden ecosystem.

In this book I will explore the spectrum of rewilding and wildlife gardening so that you can find your own balance. You can create a beautiful landscape that will maximize the impact on species who live in your garden, and even invite some in who may not yet have made a home there. You can explore all the options to find your level of wilderness, from a tiny patch to a whole plot. If you're new to gardening or are finding garden maintenance overwhelming, it's lovely to have a planned retreat and take the pressure off yourself a little.

No matter how big or how small your green space is, there is something positive you can do to support threatened species and create a rich habitat. If we understand how one plant or creature supports another, we can work out how we can help in our own small way and, in our own small patches, we can make a big difference.

5 REASONS TO REWILD

1. TO CREATE A NATURAL AND UNSPOILT HAVEN FOR WILDLIFE right outside your door. By relinquishing a little control on our gardens, we can create a safe and friendly environment to welcome the wildlife in.

2. TO REDUCE CHEMICAL USE, which has wreaked havoc in our gardens for decades. We feed plants and lawns with chemicals, we kill weeds with them, we poison rats, slugs and insects with them, we remove fungus with them and we use them to burn and to clean with. But the application of chemicals can disrupt a whole swathe of our flora and fauna. Rethinking the way we apply chemicals, how many, what kinds, when to do it and when not to do it, with the aim of avoiding their use altogether, will make a huge difference to the health of our natural landscapes, both in and around our gardens.

3. TO CHERISH UNDERVALUED PLANTS SPECIES. We have deemed certain plants as 'weeds' when, in fact, many have hugely beneficial qualities both for us and for visiting wildlife. Many so-called weeds are actually native wildflowers with a rich local history that can be a valuable food source for wildlife and even sometimes for us. They can be used for egg laying, nesting and nourishing large numbers of species. Allowing some to get a footing can be both beautiful and useful.

4. TO BALANCE THE ECOSYSTEM IN OUR GARDENS. Some insects, birds and mammals have historically been evicted from our gardens and even trapped or killed. To resort to such destruction of an entire group of organisms without consideration for the wider impacts on other species can be very damaging to the ecosystem around our gardens. Letting nature find a little more balance by trusting that with more pests, there will be more predators and so on, will allow natural ecosystems to recover a little in our local areas.

5. TO PROTECT OUR PLANET'S FUTURE. Wildlife is essential to our survival as a species, which is inextricably linked to all others – though we may sometimes have lost sight of that. Our gardens are not sterile environments, curated solely for our pleasure and their visual merits, but have the potential to be bustling, busy and beautiful homes for a host of species – fungi, animals, insects, lichens and plants alike. This web of species that feed each other, house each other and rely on each other is delicate and easily interrupted. Rewilding our gardens to any degree will help to rebuild and protect those links, and provide a haven for creatures, plants and many more species, some of whom we aren't even aware of and yet who may prove to be the key to our planet's survival.

YOUR GARDEN ECOSYSTEM

Whether you've designed it that way or not, your garden is a living, breathing ecosystem. In this chapter we will look at how that ecosystem is broken down, how each individual layer works in isolation and as a whole functioning tapestry. The aim is to understand how you can create a diverse environment in your own space, filled with all kinds of life and especially with plants, which are often overlooked but both readily available and crucial in wildlife gardening. They are the backbone of our gardens – the nervous system that connects and enables other life forces to thrive. If you bring in the plants, the rest will follow.

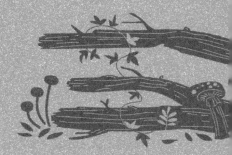

THE SOIL

Every ecosystem starts below the ground. Our soil is really an ecosystem in itself, filled with microbes, invertebrates, insects, worms, nematodes (tiny worm-like species), larvae of all kinds, roots, water, air, fungi (mainly as mycorrhiza that aids plant roots), algae, spores, different minerals and molecules and so much more.

Essentially soil is made up of three types of inorganic particles; clay, sand and silt. Clay has the smallest particles and sand has the largest, silt lies in the middle. The ratios of these different particles will determine the soil type you have and usually there will be a combination of two or three of these components. For example, if you can squeeze soil into a sausage shape and can then turn that sausage into a doughnut, then you have a predominantly clay soil. Clay soils tend to be richer and more moist but can be prone to waterlogging. Sandy soils will not bind together when wet and tend to be dryer, more free-draining, poorer in terms of mineral content but more consistent and not waterlogged in winter. Silt soils tend to be fertile and light but retain water and are easily compacted.

SOIL REWILDING

1. TRADITIONAL APPROACH – dig in organic matter to improve the soil: compost, leaf mulch or manure, unless it's waterlogged in which case either leave it or potentially add some grit or washed sand.

2. WILDLIFE GARDENING APPROACH – don't dig but lay organic matter on the surface of the soil as a mulch, allowing the natural structure of the soil to be maintained, not by you, but by the myriad species that live within the soil ecosystem. They will slowly bring the nutritious mulch down into the soil and to the roots of plants without interrupting their own habitat and cycles.

3. FULL REWILDING APPROACH – leave the soil as it is and allow the species that naturally occur in that kind of soil to grow: a survival of the fittest where any compost is formed from the materials that naturally fall onto the ground, such as leaves and twigs.

UNSUNG HERO: BACTERIA

Bacteria are incomprehensively vast, both in terms of the number of species and individuals. They are invisible to the naked eye and, in some cases, even to our most powerful microscopes. We are only just on the tip of discovering what some of these incredible organisms do and what a hugely important role they play, in the landscape, plant and animal species and within our own bodies.

GROUND LAYER

This is the ground level including all the leaf litter and natural detritus that can be found on the forest floor and provides a fantastic habitat for insects of all kinds and fungi that feed on dead wood. Here you'll find ferns, mushrooms (the fruiting bodies of fungi), ground cover species, liverworts, mosses and many, many more organisms. This is the area where small mammals can move safely from place to place concealed in the undergrowth and feeding on the worms and insects that they find in the leaf litter. Birds such as thrushes, blackbirds and robins are also found here, digging for small invertebrates to eat.

GROUND LAYER REWILDING

1. TRADITIONAL APPROACH – this layer should be free from weed seedlings, leaves, twigs and bark fallen from trees. The ground should be raked so that you can see the soil and lightly forked to give a fluffy finish. This approach may look neat but it is totally unnecessary. This is a vital layer of the ecosystem, housing all kinds of insects at all stages of their development. These insects are needed to become predators for other insects that you will want to be rid of during the growing season. In raking and clearing the ground layer you are removing a whole food source for more desirable garden visitors such as field mice, hedgehogs and garden birds.

2. WILDLIFE GARDENING APPROACH – remove some of the more tenacious weed species but make a minimal impact. Let the natural species be and leave leaf litter and small twigs to build up.

3. FULL REWILDING APPROACH – leave this layer exactly as you find it. Let the weeds grow and leave the dead wood where it is; let nature take its course. This is much less work, of course, and the area will eventually look reclaimed. If you have a big enough space in the garden, you could certainly adopt this method for some parts. If you're really keen, you could do this for the whole space but be warned that it may appear messy to some.

UNSUNG HERO:
BEETLES

One of the most effective predators of pests in the garden, beetles – especially the ground beetle species that doesn't fly – are entirely dependent on the ground layer level. Beetles live here, walk safely here and feed here. While these species are not glamorous, some are beautiful and they are hugely important to biodiversity as they control the numbers of other insect species, significantly reducing our need to turn to harmful pesticides, which devastate the entire insect population.

UNDERSTOREY LAYER

These are the bulbs, grasses and herbaceous perennials that grace our flower beds. In a natural woodland setting it would be the bracken, ferns, bluebells and grasses but in our gardens, especially in recent years, this is where we find the most flower power. Prairie planting schemes, which have become so popular in recent decades, rely entirely on this understorey layer as they are made up exclusively of herbaceous perennials and grasses. There is a lot of merit in this layer when it comes to wildlife. There is plenty of food, plenty of material for nest building, plenty of shelter from the sun and places to hide from the prying eyes of cats and birds of prey. But the inherent problem with this layer is that it is seasonal: from the bulbs of the spring, through the summer and into the autumn there will be some plant material, but the herbaceous layer will die back in the winter, leaving bare earth, if there is no ground-cover planting. However, as part of the whole multi-layered ecosystem it is integral and possibly one of the most species-rich layers.

UNDERSTOREY LAYER REWILDING

1. TRADITIONAL APPROACH – cut things back in the autumn, once the leaves and top growth has turned brown. Unless you live somewhere very cold, in which case you'll let the top growth stay on and protect the crown of the plant during the cold wintry months and cut it back in the early spring.

2. WILDLIFE GARDENING APPROACH – cut back your herbaceous perennials in the spring, maximizing their use for the local wildlife in the winter. Seed heads look beautiful covered in frost and catching the low winter sun, too! You can weed some of the most persistent invaders, such as dandelion, milk thistle, bindweed, ground elder and buttercup, and clear the congestion to allow the survival of your prized specimens if they're getting outcompeted.

3. FULL REWILDING APPROACH – let nature take its course and leave this layer totally unkempt and uncontrolled. If planting schemes are full and everything spaced evenly then things should keep on a fairly even keel. You'll find that wildlife will thank you no end for the extra seeds during the winter, and for the extra material for nesting and furnishing their hibernation holes with plenty of dried grass and leaves. If you can cope with the mess and put up with a few plant casualties that won't survive the congestion, and of course some weeds popping up in the precious few gaps, then this is the easiest option by far!

SHRUB LAYER

We tend to think of shrubs as being outdated. They were a fad of the 1970s, when miniature conifer gardens were considered the height of sophistication. Since then they have become a little passé and *Rhododendron* (rhododendrons), *Camellia* (camellias) and dreaded *Aucuba* (spotted laurel) have fallen from favour. However, in a natural ecosystem, and therefore in our garden, they can provide a safe haven for wildlife. They make up our hedgerows, which can act not only as protection for our gardens, but also as vital corridors allowing small and vulnerable species to move around and between different gardens. Shrubs also provide valuable nesting sites for garden birds. So allowing space for some shrubs is really important in a balanced ecosystem.

SHRUB LAYER REWILDING

1. TRADITIONAL APPROACH – keep shrubs in a specific border. This does not always allow for a good balance of wildlife and means that birds who live in the shrubs need to fly out from the safety and cover of this habitat into the open to feed and gather nesting material from the flower borders. This is dangerous for any small bird. Shrubs are often clipped into shape and pruned regularly to stop them getting too big and unruly. Doing this during nesting season is hugely disruptive to birds and is even prohibited by law in some countries.

2. WILDLIFE GARDENING APPROACH – to create a wilder but still more measured effect, carefully choose shrub species that will provide blossom and berries for wildlife but never outgrow their space or exclude too much light. This will ensure they can live in harmony with the plants beneath them. If you feel that things are getting too large, you can carefully and sensitively prune to thin out the growth or reduce the size. Just make sure any pruning work is carried out between the end of the summer and early spring, when the birds aren't nesting, and any logs are kept and piled up as natural insect houses and places where valuable fungi, lichen and mosses can grow. This will add to the ground layer (see page 17).

3. FULL REWILDING APPROACH – let shrubs be and allow them to grow. Many shrubs have a determinate size, so if you choose your species carefully they should never outgrow their space. A variety of species that produce flowers and fruits at different times of the year are best for wildlife. With careful planning, you will end up with a beautiful space. If any branches fall off, then the dead wood provides a great habitat for fungi and invertebrates. If you let nature take its course and shrub seedlings fall then you'll find very limited species actually land in your garden and germinate. Typical seedling shrubs would be elder, hawthorn, blackthorn, some native roses like the dog rose (*Rosa canina*), gelder rose or spindle. These are all great species for wildlife but guaranteeing that they flourish can be tricky and some grow very large if unchecked.

CANOPY LAYER

Just because we can't see this layer of the garden does not mean that it isn't hugely important. I always remember my fascination with watching wildlife documentaries where ecologists and naturalists ascend the canopy of a deep jungle in south America or Asia and reach a secret, hidden world, bathed in sunlight, open and beautiful with uninterrupted views and a sense of the space – an entirely new perspective. This is what the wildlife finds up there too. Our raptor species, such as buzzards, red kites and hawks, rely on this layer both to nest and to seek prey. Tree species can also offer a huge number of flowers that provide food for bees and other pollinators, as well as prolific fruit and seeds that can feed birds, mice and even flies and wasps – all incredibly valuable species.

. . . .

UNSUNG HERO: WASP

I remember failing a job interview because I recommended a cherry tree for a small garden. 'But that would bring in the wasps!' the interviewer said. Well, as far as I'm concerned, the more the merrier. There are all kinds of wasps, including gall wasps, tiny parasitic wasps, hornets and many others. As a matter of fact, there are over 200,000 species of wasp around the world and the UK alone is home to around 9,000. Wasps perform an essential function in our gardens and in the world at large. They are crucial for decomposition and consume a lot of compostables. They are also voracious killers of other insects. This sounds bad I know, but without wasps we would have a surplus of other insects and spiders. In fact, wasps eat greenfly and caterpillars, some of the biggest garden pests, and so are hugely helpful to us gardeners.

. . . .

CANOPY LAYER REWILDING

1. TRADITIONAL APPROACH – the traditional gardener is wary of trees as they can grow so large that they take over a small space. So, in a traditional approach, species tend to be chosen very carefully – we have learnt from the mistakes of the past to avoid planting miniature conifers that quickly grow into huge beasts.

2. WILDLIFE GARDENING APPROACH – think carefully about the species you plant. Once a tree like a sycamore is mature it will seed itself everywhere, and where one is great, any more than that is too many, so regularly pull up any saplings you find. If you have a garden large enough for a sycamore, it might be better to choose a species such as an oak, a walnut or a beech, rather than letting trees seed themselves. Remember that the more varied species of plant you have, the more varied species of fungi, lichen, invertebrates, birds and mammals you will have. The effect trickles up the food chain. Choose your trees according to the natural conditions of your garden: look at soil type and climate type; visit local woodlands and see what species thrive there, that will give you a clue as to what you can grow. In a big space, larger trees will do well; in a small space one big tree and then a few small ones will work better. Do note that small species such as rowan and cherry can also provide a huge amount of food for pollinators and birds. Bigger is not always necessarily better.

3. FULL REWILDING APPROACH – if you let your garden be, you will find that some tree species seed themselves readily. Some of the most common are sycamore, which will turn into huge trees, quick growing at first but slowing down after some decades and, if allowed to grow large, quite beautiful. Yew is another that self-seeds easily, as do birch, oak, ash, alder, hazel and holm oak, especially by the coast where little else grows. If you leave these unchecked you could be lucky, resulting in a beautiful patch of native woodland right outside your door. In years (and I mean hundreds if not thousands) the species will have diversified and you will find a rich woodland. In reality, there will probably be a scrubby look, too densely planted and with few species, at least for some years.

UNSUNG HERO: KESTREL

Many of us know the plight of the barn owl with its population in decline, largely due to the loss of its natural nesting habitat in old barns, but what you may not realise is that kestrels (*Falco tinnunculus*) also nest in buildings. They do not build their own nests but use other birds' disused nests on cliffs or in buildings, or they will also happily accept man-made nest boxes. Once they've adopted a nest, kestrels can revisit it for decades to come. Usually they breed in April and May but if food is low then a kestrel may not produce any eggs at all. If you're looking to give this species a helping hand then think about adding a nesting box on your house. You can also help by trying to increase numbers of small mammals in your garden, such as voles and mice – increase the biodiversity of plants and insects, put out seeds for mammals as well as birds, or grow grains like millet, corn and wheat, and, crucially, create 'safe' spaces free from cats and dogs.

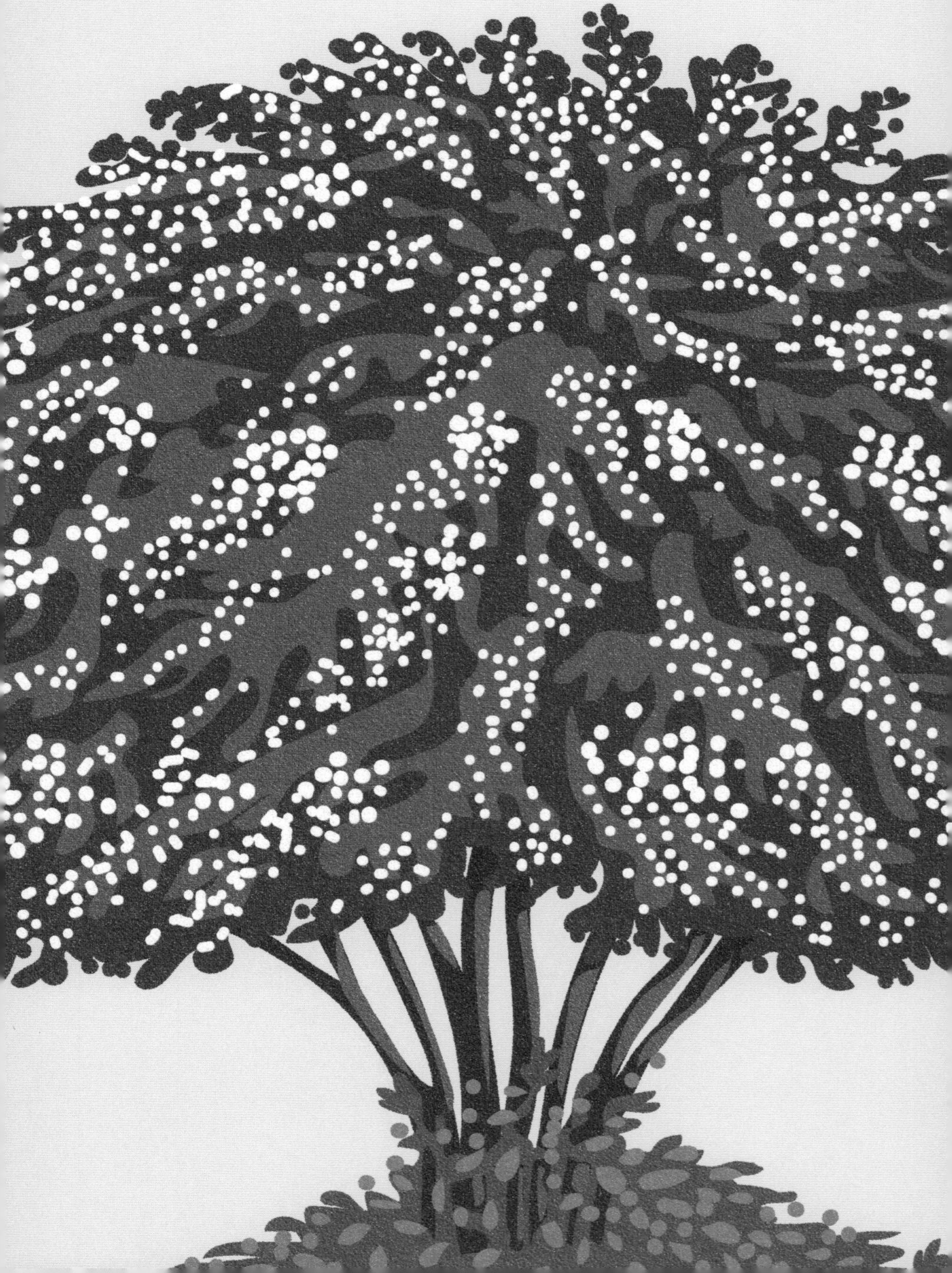

UNSUNG HERO: HAWTHORN

We see the beautiful hawthorn (*Crataegus monogyna*) adorning our roadsides in the spring and tend not to look twice at it, so ubiquitous has it become. But I love them and so does the wildlife. When in bloom, for me they truly signal spring at its best and in recent years we seem to have had exceptional displays of May blossom that fill the air with its distinctive scent (though it's not to everyone's taste!). This blossom provides a huge amount of nectar and pollen to insects in spring and its red berries are some of the last to cling on through the autumn, providing a late harvest for birds and mice. While we may not like this tree's lethal thorns, for a bird this is a safe haven where large predators dare not venture, making it the perfect home for them.

CLIMBING LAYER

It is worth noting that although this layer can span the different storeys, climbers offer a huge amount of both habitat and food source for wildlife. Climbers come in many different forms – evergreen and deciduous, ones that twine, ones that use suckers to grow and ones that grip with sharp thorns. They can flower in winter, spring, summer and autumn and some offer exceptional seed heads. Some grow in sun and others in shade. They also allow gardeners and wildlife to make the most of that bare space on the wall or in forest gardening systems, on the stem of a tree (see page 39). And, just as these plants themselves climb, so do they allow other species to access the canopy layer.

CLIMBING LAYER REWILDING

1. TRADITIONAL APPROACH – encourage climbers on walls and bare surfaces up a framework and keep them in check with regular pruning.

2. WILDLIFE GARDENING APPROACH – allow some native climbers but also plant some additional and potentially more unusual species that provide evergreen cover and some flower for pollinators. Plants such as *Trachelospermum* (star jasmine), *Jasminum* (jasmine), *Hydrangea petiolaris* (climbing hydrangea), *Passiflora* (passion flower), *Clematis* (and note that old man's beard is a form of Clematis), *Solanum jasminoides* (potato vine), *Rosa spp* (rose) and many others, provide all kinds of habitats and food sources.

3. FULL REWILDING APPROACH – ivy and old man's beard would probably be the main climbing species that would naturally occur. These are both hugely beneficial species and worth their weight in gold come the autumn when the flowers and evergreen foliage of the ivy provide nectar and shelter, and the seeds of old man's beard provide valuable nesting material.

FOREST GARDENING

In forest gardening, an increasingly popular practice, a multi-storey approach is used to provide food and flower at every level. In this process, you would grow climbers up all of the trees in your garden to maximize the space. If you choose species that produce fruit at every step of the way, you can also have a really easy, self-sustaining vegetable garden in a forest garden. Examples of a typical edible forest garden are:

- trees, such as an apple, cherry, fig or nut
- climbers like hops or beans growing up the stems
- shrub layer beneath, e.g. currants, rosemary, lemon verbena or raspberries
- artichokes, fennel, sweet cicely, lovage and borage growing at low level, providing mainly perennial but also some annual edibles
- *Gallium* (sweet woodruff), thyme, chamomile, leaf litter and twigs make up the ground layer.

UNDERSTANDING WEEDS
– OR BEING UNDERSTANDING WITH WEEDS

It's become a cliché but weeds are simply plants in the wrong place. They have found a way of proliferating efficiently, whether by seed or by spreading quickly through the soil. They come in all shapes and sizes, from the ancient fern horsetail, to the climbing invader bindweed and the huge towering sycamore. All take over and can be the bane of our lives. However, it's not all bad. Some, for example chickweed (see box on page 47) and fat hen, are incredibly useful as edibles and medicines for us. Some weeds are actually wildflowers that are in decline due to habitat loss, such as orchids and arable weeds, which rely on open grassland, and woodland-edge flowers like sweet cicely, celandine and dogtooth violet that are struggling against pesticide and herbicide use on fields.

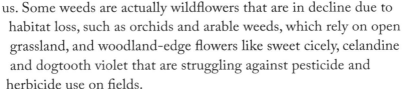

In a rewilding system, weeds are an essential part of the process. They, after all, are an ancient and integral part of our native habitats and provide a huge amount of food, nesting sites, breeding sites and shelter for our wildlife. If we want to be puritanical about it we would just leave them be, but a more measured approach would be simply to accept some weeds. Perhaps not all over and perhaps not all of them, but some throughout. A small section of your garden could be left to be completely wild and weedy, if space allows. Let weeds be a part of the process and even use them, but remove the most pernicious, the ones that outcompete everything else.

In rewilding, the whole area will eventually find a natural balance (although this will take a long time). There may be stages where your space looks completely unkempt but the slow mechanisms of nature will be at work to create an eventual harmony. There are certain species of weed though, that disrupt this balance. Invasive weeds can quickly colonize your garden and become a problem. They do not work in harmony with the native flora. They grow profoundly quickly and their reproduction is boundless. These species need to be looked out for and removed whenever they are encountered. Some are even on a list of species that need professional intervention (see Problem Weeds to Eliminate, page 44).

GARDEN FAVOURITE: COW PARSLEY

I have heard that a few years ago, cow parsley (*Anthriscus sylvestris*) was the biggest selling herbaceous perennial of the year. I don't know if that is true but, if so, I can see why. There is nothing quite like the dancing, lacy flowers of the Apiaceae (carrot) family, and in particular cow parsley that graces our road verges in the early summer. I love them and it seems like many of us are in agreement. But until recently they have been dismissed as a weed. It just goes to show that if we think a little differently we can see the beauty of plants that were hitherto overlooked.

PROBLEM WEEDS TO ELIMINATE

For any plants on this list, always get professionals in, regardless of where the plant is growing.

***HERACLEUM MANTEGAZZIANUM* (GIANT HOGWEED)** – a member of the carrot family that can grow up to five metres tall. It is easily identified by its gargantuan size, its stems covered with purple spots and its huge white flower heads. This plant is toxic and often causes severe skin irritations if you handle it, especially if you cut or damage it.

IMPATIENS GLANDULIFERA (HIMALAYAN BALSAM) – this is a
species of *Impatiens* and closely related to busy Lizzies. It has a very different nature though. Along river banks it has become a huge problem and can be easily recognized by its purple flowers, which later turn into exploding seedheads that pop as soon as you touch them, even with the lightest touch. It's precisely this that makes them such prolific reproducers.

REYNOUTRIA JAPONICA (JAPANESE KNOTWEED) – this weed may look
fairly innocuous with simple leaves and fleshy spear-like stems, but it is renowned for growing through concrete and if knotweed is known to be present it can cause a huge headache when it comes to buying or selling a home. A systemic approach using injected herbicide, only available to registered professionals, is the only way of dealing with this thug.

RHODODENDRON PONTICUM (COMMON RHODODENDRON) – this is the only species of *Rhododendron* that appears on the list. Originally, like many of these weeds, it was planted for its beauty but it has become a pest, particularly in woodland. It has proved to be the carrier (although it remains unaffected) of a number of bacterial and fungal infections that can spread through native trees, causing devastation. Any *Rhododendron* with a bright purple flower that appears, un-bidden, should be treated with extreme suspicion.

CRASSULA HELMSII (NEW ZEALAND PIGMYWEED) – this is a succulent that has travelled along waterways and caused havoc. It needs to be removed immediately, especially if you have a fresh watercourse in your garden. Never spray pesticide or any chemical near a body of water as it can cause untold damage and can flow downstream to affect a huge number of species. Always get professionals in to deal with invasives that are growing near water.

UNSUNG HERO: CHICKWEED

What a fantastic little weed this is. Firstly, chickweed (*Stellaria media*) is a great ground cover, providing a home for beetles. It's also a primary colonizer so will germinate quickly on freshly turned earth and can easily be hoed or pulled up if it's getting too big for its boots. But I urge you to cut it with scissors and eat it instead! It is delicious eaten raw or cooked as an alternative to spinach. As well as this, it has some healing qualities – it's soothing for the skin, especially if you suffer from rashes or irritations.

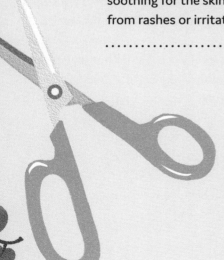

NATIVE AND NON-NATIVE SPECIES

A hotly debated topic of recent years is the value of native planting compared with non-native species. Traditional thinking is that native species are best for attracting wildlife and providing it with all that it needs. However, recent studies (particularly one undertaken at RHS Wisley Gardens in Surrey, England that looked at how pollinators reacted to native and non-native plants) have found that in reality there isn't always as marked a difference as we previously thought.

The research shows that where pollinators are concerned near-natives – that's anything from countries that surround you or are in the same biome and region – are nearly as effective for wildlife as the native species. However, when it comes to trees, it often seems that choosing species that evolved as locally to your region as possible is the most effective way of providing for wildlife. For example, an English Oak in the UK will provide a home for 284 insects alone, while in Spain the native Cork Oak woodlands provide invaluable sanctuary for the endangered Iberian Lynx and the Spanish Imperial Eagle, among others. So choosing failsafe, larger species that have a guarantee of providing habitats for wildlife is a good idea, but when it comes to smaller species the general consensus seems to be that near-native is fine, especially with a focus on flowers. The greater variety of species you have the better, as there will be more flowers for longer in the season to provide a varied diet to the local fauna.

THE CLIMATE EFFECT

Our changing climate is an additional complication for this debate. Certain animal and bird species that we used to take for granted are becoming less and less reliable visitors to our shores, whereas other species, previously unseen, are becoming more common. This is, of course, why we choose to plant for wildlife, to give our dwindling species a helping hand, but if we find ourselves hosting visitors who are a little more unexpected but in no way damaging, should we not also provide them with food and shelter? So just as climate change allows us to grow more varied things, it can also allow different wildlife to visit.

The only thing we know is that plants from the opposite hemisphere and completely non-natives, like tropical plants in a temperate climate or vice versa, will provide very little for either native or visiting wildlife. If you like the challenge of growing things that do not naturally thrive in your garden or region, then you'll have to work extra hard to bring in other species by cleverly concealing the odd native or near-native in your schemes.

KEEPING THE BALANCE

When it comes to rewilding, it can't always be assumed that the species that naturally seed themselves will be native. In fact, some of our most determined weeds are invasive species, as mentioned earlier in this chapter (see Problem Weeds to Eliminate, page 44). If you are rewilding in the purest sense and letting nature do the work, keep an eye on that balance. If you notice too many non-native species beginning to take over, it might be worth intervening a little to redress the balance so that no one thing is able to gain too much of a foothold.

FEED THE ECOSYSTEM

The answer, whether you plant native species or near-native species, is to have as much variety as you can. Build areas for native species intermingled with areas for near-natives (as well as some non-natives if you love them) with as much flower as you can get for as much of the year as you can. Make sure there is a variety in the types of plants you have – climbers, shrubs, trees, herbaceous perennials, annuals, grasses and ground cover of all different kinds, some evergreen, some deciduous. Get as many flowers as possible, both on trees and shrubs and on herbaceous perennials and annuals, and keep a healthy soil profile by allowing a build-up of leaf litter and organic matter, reducing the digging that you're doing and keeping some dead wood and hidey-holes for insects. In this way, you are keeping the garden ecosystem going at full power. The cycles of the relationships between the different species will be functioning and will create a natural and self-sustaining balance.

Whether you fully rewild or choose to intercept nature at times, to intervene and stop something happening or encourage something to happen, it all makes a huge difference. Our gardens are a functional ecosystem with layers; webs of complicated processes and relationships, most of which we are totally unaware of, happening right under our noses and under our feet, a thriving, living, breathing habitat. If we think of our gardens like this, not just as a room outside our house or a manicured space designed purely with aesthetics in mind, then we can begin to train ourselves to make the changes so badly needed by our struggling wildlife.

BRING IN
THE WILDLIFE

In this chapter we will explore the areas around your garden section by section that could be an untapped resource for you and your wildlife. We all have routines with our outdoor spaces; practices we've got used to over the years and do without thinking. But if we look closely at the patch outside our door we may begin to see things a little differently. It may be the forgotten corners that hold the unlocked potential, or it could be a huge area of our garden, so big we almost don't see it and don't realize what a haven it could be with a few, small tweaks!

WATER

Water is the basis for all life. Without it you will see very little wildlife: creatures may visit but they probably won't really interact with your garden. This is because water serves so many functions. It's a place where animals wash, drink, lay their eggs, spend a part of their life cycle and even shelter.

Water is vital to our garden ecosystem and is a great tool by which we can transform our gardens into buzzing, scurrying and thriving communities. It doesn't matter if you have water naturally occurring or not. You can control the amount of water you have, where you have it and even how it looks, allowing you to truly create this element of an ecosystem – it can be entirely artificial, but nonetheless integral.

Water in our gardens comes in many shapes and forms. Many of us wildlife lovers have a bird bath and enjoy watching it

being used on a regular basis. Birds love this feature and having it placed up high gives them a sense of safety, making them feel relaxed enough to use it without the risk of a hungry cat or dog sneaking up on them.

While a bird bath is great for birds, it's pretty useless to other animals, and often the water gets splashed about and needs topping up so regularly that it's not there for long enough to act as a nursery for waterborne insects. So, I would advise keeping the bird bath but I would also suggest adding another water source at ground level for other visitors.

What I will do is paint for you a picture of the perfect pond, one that makes the absolute most of its potential as a haven for wildlife, and you can take elements from it when designing your own. Remember while you read this though, that there is no shape or size that doesn't work.

THE PERFECT POND

Even if you are a committed rewilder and want nature to take control completely, a pond or other source of water is probably the one essential human intervention you will need to make. It is not an option, it is a minimum requirement. Without water, you will likely end up creating a haven for plant, fungus and lichen species without necessarily bringing in any wildlife.

If, at the other end of the spectrum, you want to maintain a garden that looks like a 'garden' and fits in to your ideas of aesthetics and design, and is a place you feel most peaceful when not overrun with weeds, then a pond makes a highly appealing feature – even adding trickling and splashes to the soundscape of your garden. Your garden can be modern and sleek with straight lines and minimalist planting schemes, and still benefit from the formality and focal point of a well-designed wildlife pond.

Some or most of us probably fall in-between these two stances, whether consciously or not. You might simply not have enough time to keep your garden as manicured and pristine as you'd like and there is a lovely reassurance in being able to create something formal and geometric; giving the garden some structure to balance out areas where there may be none, or echoing the sinuous forms of nature and creating a pond that is a little more informal. The design is up to you. Make the pond diminutive and almost hidden in a corner you hardly use, or make it a huge statement in the middle of the garden, and use any shape you like. These things simply don't matter. What matters is all the other features that I am about to describe in the perfect wildlife pond!

....

GARDEN FAVOURITE:
DRAGONFLY

This species relies on water for its life cycle. Adult dragonflies lay their eggs near water and once these hatch, the young, called nymphs, descend into the water depths where they consume small insects and even tadpoles, until they're large enough to climb up a stem, emerge from the water, spread their wings and fly for the first time. If you want to bring dragonflies to your pond it will need to be about 60cm (2 feet) deep in some areas, get some decent sunshine and very importantly, have reeds or stems that grow from the water up and out into the air, both for the emerging flies to take flight and for adults to sit on and attract a mate. Creating the right environment for them is worth it to see those beautiful flashes of blues, reds, yellows and greens as these majestic insects dart around our ponds.

....

POND DOS AND DON'TS

Following these basics rules will ensure you have a well-designed, useful pond in your garden that will attract all manner of wildlife.

DO...

THINK ABOUT ACCESS POINTS: there needs to be safe access in and out of the pond for all kinds of wildlife. Adding a pile of rocks that slopes up on one side gives an ideal spot for lizards to bathe in the sun when the rocks warm up or, if it's a large pond, create a shallow end and a deep end. The simple reason for this access issue is the safety of the animals that come there to drink. You'll find all sorts of creatures use a pond in this way, from hedgehogs, mice, birds, voles, wildcats (if you're lucky enough to have any), to bats if your pond is large enough with uncluttered water, and even wasps and other insects. Any non-aquatic animal who falls into a pond without an escape route will sadly drown in the water. Most animals can instinctively swim, even if they would prefer not to, so will usually swim around for a while trying to find a way out and the ideal pond will have one or two safe exit points so that, at worst, the animal is a bit damp.

In fact, this point is so important that I would recommend some stealth action in your local area to add escape routes in any large bodies of water such as ponds, canals and rivers, just in case. I once leaped into a canal to rescue a hedgehog that was swimming up and down the concrete-sided waterway trying to find a way out. He was fine once he caught his breath, but a little ramp provided every now and again would save many a furry life.

PROVIDE SHELTER: this can be anything from an overhanging rock to a glossy lily pad. All creatures that dwell in the depths of the pond need some shade as water can get pretty warm during the summer months – especially if it's shallow. Providing a shady place where insects, newts and water snails can keep cool creates a much healthier and happier population.

PLANT SOME OXYGENATORS: most plants in water are optional but one group is necessary and that is the oxygenators. They perform the vital role of (rather obviously) keeping the water oxygenated, which allows reptile, amphibian and insect species to thrive, but also keeps things like algae and phytoplankton from taking over. Oxygenators also help prevent the water from becoming stagnant. They are very easy to get hold of and some of the easiest and hardiest to add to your pond are *Elodea canadensis* (Canadian pondweed), *Ceratophyllum demersum* (rigid hornwort) and *Fontinalis antipyretica* (willow moss). These go below the surface of the water but some, such as *Isolepis cernua* (slender club-rush) or *Marsilea quadrifolia* (European water clover), grow above and below the surface of the water. Some flower, such as *Ranunculus aquatilis* (common water crowfoot), adding beauty as well as practicality to your pond. Adding a water feature with a filter and pump can also really help to keep the water clear and moving a little.

ADD SOME SAFE CORRIDORS: while this isn't strictly a part of the pond, there is a good argument for providing a safe route from any hedge to the pond, to allow species to access the water and then safely return to their burrow or nest.

The plants you choose for your pond can not only look beautiful but also be high functioning in that they will provide a safe haven for wildlife to hide in.

Adding some oxygenating plants to your pond will prevent the water becoming stagnant, stop algae taking over and help pond wildlife thrive.

Adding a shady spot for any animal
and insect visitors to the pond will
provide much needed shelter on a
sunny day and stop the water
heating up too much in the summer.

Whether in a pond, a canal or a lake, a simple
wooden ramp, a pile of stones or a shallow end
will give any struggling swimmer a chance to
get out of the water.

INTRODUCE OTHER PLANTS: they may not be strictly necessary but this is my ideal pond, and an ideal pond for wildlife. Having reeds, rushes and leaves to scurry around in, hide in and sometimes even nest in adds something extra to a wildlife pond, as well as helping it to look beautiful. Even the most modern design can benefit from a striking marginal such as *Rodgersia* (fingerleaf rodgersia) or *Equisetum japonicum* (barred horsetail) and who doesn't love a water lily or lotus? There is a functionality to these areas too: they provide a haven for wildlife to hide in. The advantage of a pond is also its danger – it attracts all kinds of creatures in their droves. Some of those creatures will be higher up in the food chain than others and adding other plants will give the more vulnerable species like field mice, dormice, dragonflies and small birds somewhere to hide from foxes, hedgehogs, cats, dogs and predatory birds such as buzzards and hawks.

DON'T...

INTRODUCE FISH: sadly, if you want to attract the most wildlife you will need to avoid any ornamental fish in your pond. Fish will eat most species they encounter in the water itself and many that float on its surface. Tadpoles, water fleas, water boatman and many more will be fair game for the fish. The idea of a wildlife pond is to welcome a whole tapestry of fauna, from snails to insects and even leeches, who in turn attract more hedgehogs, frogs and insect-eating birds, who then attract foxes, cats, birds of prey, and everything in-between for a drink, and possibly a wash and a meal. If you remove a whole spectrum from the food chain by introducing fish with voracious appetites – especially if you encourage them to grow to a disproportionately large size in a small area – then sadly this will be irreparably damaged. You will only find a fraction of the visitors you expected in your pond, plus perhaps the odd heron to polish off the fish themselves.

PLANT INVASIVE SPECIES: the natural world has seen ecological threats emerging from the ill-advised introduction of invasive species. In water, and particularly if you are lucky enough to have a natural watercourse running through your garden, this advice becomes much more relevant. If you plant something invasive in your garden then at best it's a pain and at worst it will require an expensive and harmful chemical removal job by professionals. If, however, you plant something invasive where its seed will either be carried away by visiting wildlife, or carried down stream by the current, suddenly you have a much more serious problem and whole swathes of waterways can be irrevocably taken over by species who inherently outcompete all the native flora.

GARDEN FAVOURITE: FROG

Although frogs are amphibians and therefore need water, they are also very regularly trapped in water that they can't escape from. Even for water-dwelling beasties like this, that spells disaster. Just like mammals and birds, a frog needs an escape route from a pond so that it can happily continue to eat all those slugs, snails and insects that we so often consider our garden foe. If you want to add frogs to your pond, it's best to get hold of the frogspawn rather than the frogs themselves, who will likely be upset at being moved and return to where they were found.

Rodgersia pinnat 'Superba'

UNDERSTANDING POND PLANTS

Plants are an essential component of a wildlife pond, to provide a variety of habitats, shade, food and some hiding places in the water, as well as keeping things fresh, avoiding stagnation and algal blooms. For the healthiest ponds, a mix of the following different plants is by far the best and most beautiful, and ideally you should aim for one-third of the water to be filled with oxygenators and half to two-thirds of the water covered with floating leaves. This will provide plenty of habitat for wildlife and prevent the build up of algae and weeds. Marginal plants (see page 66) will then provide great hiding and nestling places for water voles and shrews, as well as wetland birds.

These three key types of plant will help maintain a healthy and productive pond.

1. BOG PLANTS sit in the areas around the edge of a pond that are permanently moist but not waterlogged. If your pond doesn't have an area like this, you can easily create a bog garden next to it (see How to Build a Pond, pages 69–72). Bog plants include a host of species that are often incredibly dramatic to look at, such as *Darmera peltata* (umbrella plant), *Ligularia* (all kinds of bright and beautiful species), *Rodgersia* (fingerleaf) and *Rheum palmatum* (ornamental rhubarb).

Iris versicolor

These species all share dramatic, large foliage that looks as if it belongs to the Jurassic era, but they tend to fight for their space, so plant with a little caution. There are also more delicate-looking species of bog plants: *Astilbe* (false goat's beard), *Iris sibirica* (siberian iris), *Filipendula* (meadow sweet), *Persicaria* (common bistort) and *Primula candelabra* (candelabra primrose). Finally, there are architectural plants with minimalist foliage and few or no flowers, which lend themselves to more modern design like *Juncus* (rushes), *Equisetum hyemale* (rough horsetail) and water-loving ferns.

2. MARGINAL PLANTS sit around the very edge of a pond in the shallow water. They have permanently moist roots and, although they prefer to grow in a little soil, you can keep them in the specialist water pots they come in, if you fear the plants taking over. Some of the most notable and attractive species include *Calla palustris* (water arum), which may need some winter protection, *Carex elata* (tussock sedge), *Cyperus involucratus* (umbrella plant), *Houttuynia* (chameleon plant), *Iris versicolor* (blue flag), *Eriophorum angustifolium* (cotton grass) and *Caltha palustris* (marsh marigold).

GARDEN FAVOURITE: NEWT

Newts are voracious predators of insects, so the more wildlife we encourage into our gardens in general, the more newts we will see, but they need a pond to live in. They simply cannot survive without water, but you will find they will move in very quickly once one is built. Some newts, like the great crested, are so rare that they are in fact protected species. That means that once you have them they can't be removed. In general, newts like a clean pond so plenty of oxygenators and plants to filter the water will encourage them. Most newts, like most amphibians, will feed on land, although they lay their eggs in the water, so having a slope or rocks that enable them to leave the pond is essential.

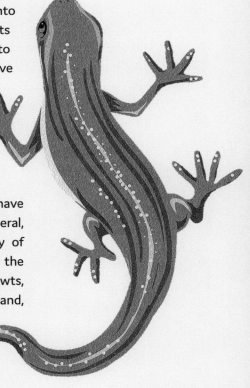

3. DEEP WATER AQUATICS, oxygenators and floating species can live right out in the depths of the water. This includes aquatic plants such as ever-popular water lilies (*Nymphaea*), *Aponogeton* (water hawthorn), *Iris pseudacorus* (yellow flag) and *Marsilea mutica* (water clover). It also includes oxygenators like *Myriophyllum verticillatum* (whorled water milfoil) and *Hippuris vulgaris* (mare's tail) that are essential for a healthy pond and floating plants like Starwort (*Callitriche stagnalis*), *Pistia* (water lettuce) and *Eichhornia* (water hyacinth, non-hardy). Although some of the more ornate deep-water species are generally categorized under one heading, always read the label before you buy them, as they can have specific depth requirements.

Equally, some can happily grow as marginals, bog plants or deep water aquatics, depending on where you plant them.

Marsilea mutica

HOW TO BUILD A POND

There are a few different methods to creating a pond and your budget a nd soil type will often dictate which one you choose to go for. A pond can be an outstanding feature in any style and size of garden, while raised ponds or ponds in containers can also look stunning. Here, I will describe how to build a classic ground-level pond, simply because that is the most useful to wildlife who are unable to access the water on a high-sided pond.

Start by working out what shape of pond you want. If you want something formal, such as a circle or a square, then a preformed pond liner might be your best option (they do also come in less formal shapes, too). They are fairly cheap but do not reliably cater for the 'shallow end' mentioned earlier (see Pond Dos and Don'ts, page 58), although they often have shelving to allow planting at different depths.

If you want something less formal or especially large there are a few options open to you. If you have thick clay soil that doesn't drain at all (great for ponds, but awful for planting in!) then you won't need a liner. If, like most of us, you have a soil that does drain away, then you'll need some kind of liner. This could be made from butyl, concrete, bentonite or puddled clay – each option has an incurred cost and maintenance implication. Some can be altered at a later date, like the clay or the butyl, whereas materials like concrete are pretty much there forever. Most of us will use a liner as it's cost effective and versatile, so I will describe this method.

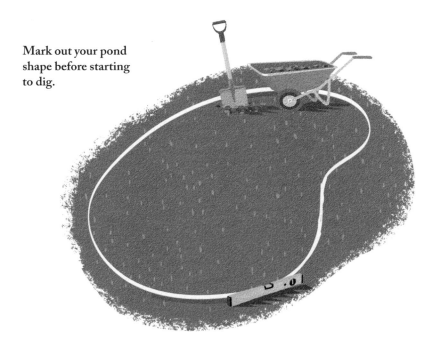

Mark out your pond
shape before starting
to dig.

1. MARK OUT THE POND SHAPE using spray paint or string on the ground. A pond should be placed near enough to trees and hedges that the creatures nesting in them can safely get to it, but it should not be directly under trees as the leaves of broadleaf trees and needles of conifers will fall into the water and create a build-up of sludge in the bottom. As this rots, it will increase the acidity level of the water. You should also position your pond somewhere where it will get some sun.

2. START DIGGING OUT the shape you want, making sure there is at least one shallow end so that wildlife can climb out of the water more easily. Ideally you should get to a depth of at least 60cm (2 feet) at the deepest point, the deeper the better. Always make sure your hole is level – if one side is higher than the other, you will find the water pours out on the lower edge and you'll always have liner visible at the higher edge.

3. REMOVE ANY STONES from the hole and use an underlay to line the soil before adding a layer of sand.

Lay a pond liner over an underlay.

4. LAY YOUR POND LINER over the top of the sand, leaving a large overhang around the edges.

5. START FILLING THE POND with water. Ideally this should be rain water but it could be tap water that is left for a few days before you plant. As the water fills up, the liner will be pinned against the sand. Keep an eye out as this happens to make sure there are no folds in the liner. If there are, smooth it out as the water fills up.

Fill the pond, smoothing the liner as you go.

6. ONCE THE POND IS FULL, place stones over the edges of the liner to hold it in place and hide it, then trim off any excess.

7. BEGIN ADDING A LITTLE SOIL or compost around the edges of the pond and adding plants. This will initially make your water cloudy, but don't worry: it will settle and clear completely in a few days or weeks.

8. IF YOU HAVE THE INCLINATION, make a bog garden by burying a little excess liner under a flower bed next to the pond. You will need to make a few small holes in the liner before you bury it to allow a little drainage to your bog plants. Cover the liner with soil and plant it up with bog species (see Understanding Pond Plants, pages 65–68). You can expand this feature in later years if you like.

Plant a bog garden at the edge of your pond.

MAINTAINING YOUR POND PLANTS

This process is really important because by their very nature pond plants tend to spread and proliferate until there is too little water left visible. There are no hard and fast rules but, generally speaking, there should be at least one-third of the water's surface visible. This will help wildlife find your pond. A clear-out every few years should be enough to stop any one species taking over and keep a good water to plant ratio.

It is also criticial to remember that most of the species living in your pond will be extremely small (a regular pond dip will reveal to you just how teeming with life your pond is) and they tend to live, or at least spend a lot of their time, in the plant material below the surface. So, it's imperative that after you have removed any unwanted plants and material, you leave it on the ground immediately next to the pond for a couple of days, to give all of those little creatures a chance to slip safely back into the water.

Ponds can be prone to pernicious plants such as duckweed and blanket weed. Although some low levels are completely normal and beneficial, once they are clearly visible from the surface, you have a problem. To help reduce the risk of these taking over make sure you keep your pond well balanced with oxygenators (see Pond Dos and Don'ts, page 59). Add material like barley straw to the water to reduce the blanket weed if it becomes too overpowering. And a good old-fashioned fishing net will bring up most of the blanket weed and duckweed, which floats but doesn't go down deep in the water. As before though, remember to leave discarded weeds next to the pond to allow critters to return home.

Top up water levels in your pond with rain water in summer to stop the habitat drying out and in winter regularly break any ice on the surface so that oxygen levels in the water don't become too high.

LAWNS

In the following pages it may sound as if I have a vendetta against the lawn, and perhaps in some ways I do. However, for those lawn devotees out there (and there are many), a lawn is infinitely better for the environment and our wildlife than many other options, including the dreaded astroturf, paving, decking and even gravel. So, if you are an advocate of the perfect patch of Poaceae (that's the grass family) then you're doing a great job. There are, however, a few things about our fascination with a finely mowed turf that leave a little to be desired.

WHY GRASS?

Well, it's soft, it isn't spiky, it's green, lush and we can walk on it. That is the short answer. It serves its function as ground cover very effectively. But there is an unhealthy preoccupation with grass in particular in my opinion. As anyone who has ever tried to maintain a lawn will know, there is always a ready colony of other species waiting to slowly take over little patches of our lawns. When you think about it, these species such as plantain (see box, page 77), daisy and clover are perfectly adapted to perform exactly the same function as grass – they are low growing, can grow even when we walk or run on them and are not spiky.

So why do we give preferential treatment to grass? The more I think about it (and I have given it a lot of thought over the years) the less I understand why we encourage grass and eradicate anything else. We wouldn't do it in a woodland or a flower bed, we would instead aim for a nice mix of different species. But when it comes to our lawn we have species blindness – even to the point of using strong chemicals to obliterate anything that's not strictly grass. In general, species diversity is what encourages the optimal conditions for the most wildlife. A slight tweak in the way we think of our lawns could encourage hugely diverse wildlife and wildflower species and have really helpful environmental effects with very little effort from us.

ADDING DIVERSITY

Whatever your preference and desired aesthetic, and also whatever size your garden may be, having a little patch or even a pot of lawn grass can provide huge benefits for the wildlife. Domestic animals like to eat it when they feel unwell, insects hide in it, pupate and grow in it, lay their eggs in it and hunt and move safely in it. Birds use all kinds of grass to make their nests and feed on the seeds it produces, mice nest in it and feed on it, and we love to look at it, lie in it and play in it.

The obsession we have had for centuries about getting the perfect, green, carpet-like finish has long baffled me. I personally don't like the look of it but also, and more importantly, it offers very little to nature. A monoculture of grass, manicured, quite literally within an inch of its life, and regularly doused with chemicals is never as beneficial as a species-rich patch of grass or meadow. However, a lawn that has been left without chemicals can provide an hospitable place for invertebrates and animals to make their homes. Even better, let the grass grow a little longer, leave it completely or transform it into a wildflower meadow (see How to Plant a Wildflower Meadow, pages 88–90).

There is no point denying that grass is a tremendously important part of our gardens. We value it. But we must encourage it to be a hugely valuable resource, too; a major component of our garden ecosystem that, with a few little tweaks, can play a vastly important role in protecting our wildlife and provide a potential home for rarer wildflowers.

UNSUNG HERO: PLANTAIN

Plantain (*Plantago major* and *Plantago lanceolata*) is a weed we have hitherto overlooked but in the last few years it has had a small resurgence in popularity based on aesthetics. It's a dainty plant with little whorls of tiny cream flowers perched atop long, immeasurably delicate stalks. Look back into our history and you will learn that humans have long valued this plant for altogether more practical reasons; as food. The grains running up the stem of greater plantain, and even the seeds on the more delicate and ornate species, are edible and incredibly nutritious. They have a nutty flavour and can be soaked and eaten (they sometimes have a slightly tough seed casing) or used in cooking. The leaves of narrowleaf plantain, which are long and strappy with parallel veining, can also be used as a poultice and plaster for minor wounds. It's a beautiful, useful and delicious plant that I think deserves a little more respect, if you please!

RETHINKING LAWN CHEMICALS

Maintaining our lawns has a high carbon footprint and is chemically intensive. We feed them, weed them (often with herbicide) and we mow them (often with petrol mowers). As we are all becoming acutely aware of the damage we are doing to the environment with our everyday actions, and many of us are trying to make those little changes that will make a small difference, I have personally adopted an almost entirely chemical-free lawn care approach, especially as it's the first rule of thumb when trying to encourage wildlife in any area of the garden. Spraying for one pest or weed will invariably kill all the others and a good chunk of the wildlife to boot as the chemicals make their way up the food chain.

LAWN FOOD

I now never feed the lawn, as it encourages the grass to grow at the expense of other plants that make a more species-rich habitat and more diverse food source for wildlife. Chemical lawn feed also seeps into the ground, causing damage to worm, bacterial, fungal and invertebrate species that live within the soil and make up a hugely beneficial portion of our garden community.

HERBICIDE

I also never use broadleaf herbicide on a lawn. Any weeds that cause particular offence and take up room can be removed by hand with ease. You can regularly pull up sycamore saplings and nettles, as well as dig out the odd bramble with relative ease and absolutely no chemicals. The only reason I can see for using chemicals would be the presence of an illegal invasive species or an invasive species that will cause a real imbalance in the ecosystem. In these instances it's always better, and sometimes even required, to call in professionals to help you deal with the problem. Using strong chemicals to deal with the odd daisy, plantain or dead nettle (all of which are fantastically beneficial to insects) really is overkill, in my opinion.

PESTICIDE

Some pests such as leatherjackets, chafer grubs and other larvae live under the lawn and can cause a little damage to the roots of plants. I very strongly feel though, as hopefully anyone reading a book about wildlife will understand, that to encourage some visitors, you have to accept all. Yes, you might choose to draw the line at a hedge full of stinging nettles for example, and so have fewer butterflies, but that's not quite the same as using a pesticide to kill one particular species, which actually ends up killing many more. The simple truth is that there are no chemicals that target just one species of insect. They kill indiscriminately and without insects, you will not have many birds, mice, hedgehogs and anything else further up the food chain. Or, worse still, those animals that consume poisoned insects will be poisoned themselves. If something is causing your lawn desperate destruction, then there are certain biological controls – usually nematodes (microscopic worms) or parasitic wasps – that target one particular species. But the best approach is to encourage lots of every species in the garden, including predatory insects and birds as well as companion plants of all kinds, and then you'll find that nature is very good at finding its own balance and keeping numbers on an even keel.

MOWING

The only chemical I use on a lawn is a little two-stroke fuel for strimming (although I have just invested in an excellent battery powered strimmer) and petrol for mowing. I have a push mower that works mechanically without requiring any fuel and it really does work brilliantly for any flat surface. It even has a roller on the back if I want to create stripes. The only problem is that a lot of gardens don't always have a flat surface. So sometimes, you need a different alternative. If you have a domestic garden then an electric mower should work perfectly. But if you are lucky enough to have a huge space and you want a lawn that is well-kept, a petrol mower is probably the only option at the moment, until electric models are available. In this case, perhaps consider reducing the frequency of your mowing or the method – only mowing pathways rather than the whole lawn, for example – to reduce your carbon footprint.

GARDEN FAVOURITE: BUTTERFLY

Who doesn't love a butterfly? These beautiful, colourful insects with a delicate appearance have been championed in recent decades as a species that needs our help. The key to doing this successfully is all about your local area. There are many groups dedicated to butterfly conservation, so contacting a group local to you to find out what species occur in your area and the specific plants they rely on will help you when choosing what to plant. Each different butterfly needs different plants to complete its life cycle. A few good general choices though, are stinging nettles, buddleja and hops. Wildflower meadows will also attract huge numbers, though again, the exact species your meadow should feature will vary from region to region. Something we can all do is try not to destroy caterpillars, which are, of course, young butterflies and moths. Some may cause small holes in our plants, and even decimate our crops – particularly brassicas – but they will grow into beautiful butterflies, so if we can move or better still, ignore them, we will be doing our bit for the butterfly population. Get to know your caterpillars as well as your butterfly species so that you can monitor and document their success year after year.

THE HIDDEN BENEFITS OF MOSS

It sounds like one plant: moss. One common enemy (if you like a nice neat lawn, that is). In actual fact there are lots of species of moss and many perform a vital role in the world's ecosystem. These are some of the most ancient species of plant, growing even before the ferns. One such moss, *Lycopodium,* was among the first plants to have evolved and, back in the Carboniferous period (approximately 360 million years ago), would have grown to gargantuan size, making forest-like ecosystems. Today, this species, also known as clubmoss, is tiny and has healing qualities, which are often exploited in alternative medicines. But that's just one kind of moss. There is also cushion moss, tamarisk moss, forklet moss and the *Sphagnum* species known as peat moss, which are the species responsible for the slow formation of peat bogs over thousands of years. There are also hornworts and liverworts, which are plant species closely related to moss in evolutionary terms and not unlike moss in appearance, but quite different in quality. Although all tend to grow in damp, shady areas, and are good at colonizing bare ground, the fantastic ability of moss to absorb, is what makes them so unique and so precious. Moss can absorb water and is also extremely efficient at absorbing huge amounts of CO_2 from the atmosphere. So, every time you rake up moss, you are also reducing your lawn's ability to improve air quality.

Moss can be a very useful tool for wildlife, too. Birds and mice use it to build their nests and some species of moss make a great place for insects, spiders and beetles to hide. I like the feel of a mossy lawn underfoot but even if you don't, I would recommend keeping a little patch aside where you let the moss stay.

USING PEAT IN HORTICULTURE

Peat has long been used as a fuel and growing medium. It is harvested from all over Europe, dried and often burnt. Because it is made from *Sphagnum* moss that has decomposed in a unique way over thousands of years, it is a finite resource that is extremely difficult to renew. It is also a habitat for many species that struggle once the peat has been harvested. The quality of peat is great for growing plants – it retains moisture without becoming too heavy and waterlogged and is inert, meaning growers can add whatever nutrient they need. However, as moss absorbs CO_2, so do peat bogs. When peat is harvested for industry, that CO_2 sump is released back into the atmosphere. Horticulture as an industry is responsible for around 1 per cent of peat harvested. It may not sound like a lot, but we tend to get the oldest, deepest and best-quality peat, meaning when harvested some of the largest CO_2 sumps are released and richest habitats destroyed. However, peat-free composts of all kinds are available to gardeners – from those made of sheep's wool to coconut fibres or bulrushes, as well as soil-based options. I always buy peat-free composts and encourage my local garden centres to stock a wide range of peat-free options.

GARDEN FAVOURITE: ROBIN

Most robins are fiercely territorial, bold and inquisitive, with a lovely song and a beautiful red breast, making them a firm favourite. The reason they are so bold with us, especially when we are gardening, is that we tend to provide them with delicious worms and insects – their favourite food – as we turn the soil. We tend to see them less in the summer when food is plentiful and they disappear off to the woods to find it. Infinitely resourceful birds, they can nest pretty much anywhere; inside flower pots, under tables, in garages and basically anything you've left lying around. From laying to fledging the brooding process only takes about a month. They start nesting as soon as the winter thaws and go right through the summer, so consequently can have up to five broods a year. Most robins, like many animals, struggle to make it into old age, but if they make it through the difficult first year they can live for up to 19 years.

MOWING REGIMES AND WILDFLOWER MEADOWS

Changing the way we handle our lawns can encourage wildlife into our gardens. Traditionally we mow the grass regularly throughout the growing season. We start in the spring or the late winter and mow right through until the first frosts. The only time we leave our grass alone is in the depths of winter when the garden and most of its creatures have gone to sleep. But this isn't the only way of maintaining a lawn.

Wildflower meadows have been an increasingly popular garden feature in recent decades and there is no denying they look stunning and create a much needed boost to our rural ecosystems. A huge number of species rely on wildflower meadows (including the often overlooked rare flowers themselves), but one that benefits the most is the butterfly. For basking in the sun, laying eggs and finding nectar-rich flowers, the butterfly literally has a field day in a wildflower meadow. In order to really benefit these species, there needs to be long grass or flower meadow for most of the summer.

Last summer, while walking through orchards near my house when the grass had not been cut for a few weeks and was getting long, I disturbed 20 or 30 comma butterflies who came fluttering out of the grass for every step I took. This wasn't a wildflower meadow as such, but rather a species-rich bit of grass with a few 'weeds' that had been left to grow a little longer than usual. A few days later they cut the grass and the next time I walked there I saw nothing. Not one butterfly. Whether they had been killed by the mower or just moved on to better pastures, I don't know, but I felt very sad at the loss.

Creating a wildlife haven in your lawn can be as easy as this – either reduce the cuts you do per year or leave a patch where you don't cut at all. Long grass is a great resource for all kinds of critters, whether it's to hide, feed, pupate, lay eggs or as material for nests and dens.

Here are a just a few species you can expect to help by letting a patch of lawn go wayward (there simply wouldn't be enough space in this book to list them all):

1. Frogs and froglets
2. Craneflies
3. Sawflies
4. Butterflies
5. Moths
6. Bumblebees
7. Grasshoppers
8. Hedgehogs
9. Beetles
10. Bats
11. Mining bees
12. Blackbirds
13. Swallows
14. Robins
15. Songthrush
16. Lady's slipper orchid
17. Helleborine (*Cephalanthera rubra*)

If you want to encourage these kinds of visitors, the easy solution would be to stop lawn care. Do nothing; don't apply fertilizer or herbicide in the form of strong and harmful chemicals, and restrict mowing – perhaps once or twice a year, either in early spring, high summer or autumn depending on your needs (see page 90). It couldn't be more simple and it would be incredibly effective and beneficial. If you're lucky enough to have a large garden you could leave a section unmown, or with just mown paths, which can look attractive and still allow you to have an area of crisp and carpet-like lawn. Just make sure that you leave extra room for chemical drift if you do choose to spray any of your lawn. Any chemicals that blow onto your wild grass will make it very difficult for wildlife to use that area for food, habitat or nesting material. If you have a smaller garden then make your grass count, by keeping it all long and luscious. And especially full of weeds!

UNSUNG HERO: MOTH

Rarely seen but always there, these delicate creatures are not only misunderstood but hugely underestimated. We tend to think anything beautiful in this family must be a butterfly when in fact moths can be gorgeous. Just look at the cinnabar moth or the tiger moth. But more importantly than the way they look, moths are exceptionally effective pollinators and have even been shown to distribute pollen over a far greater distance than bees. They tend to come out at night and can see white colours clearly, as well as having a good sense of smell. So anything strongly scented and white, especially if it smells more at night, like night-scented stocks and *Magnolia grandiflora*, have specifically evolved to be pollinated by moths. Plant more pale, scented plants to give the moths something to feed on when we are fast asleep!

HOW TO PLANT
A WILDFLOWER MEADOW

If you want to go the whole hog and create a patch of wildflower meadow then good for you – it is a little more interventionist than simply leaving your grass to fend for itself, but it can be truly beneficial for our wildlife.

Wildflower meadows have a reputation for being difficult to achieve and that is down to two main factors. Firstly, most wildflowers grow on poor soil with little nutritional value. Grass is high in nitrogen when it breaks down (especially if it has been regularly fed in the past by us) so your once manicured lawn may not be the easiest starting point. The other reason is that people see images of wildflower meadows and expect theirs to look a certain way. However, the reality is, just as in planting up a flower bed, the rule of 'right plant, right place' is ever-present. You may want annual poppies to feature heavily, for example, but your garden might be in shade. You may get a couple of poppies to flower with much coaxing the first year, but realistically, by the second year, other, more well-adapted, shade-loving species will have grown, meaning the poppy seeds will find it very hard to germinate and your meadow will lose its red look. The answer is to be free from expectation.

Before you begin, contact your local wildlife charities and ask them for guidance. Often local initiatives will offer seed packets or information about the local bird, bee and butterfly populations and also the kinds of wildflowers that grow well in your area, benefitting the target species. Once you're armed with that knowledge you should be on the road to success. With guidance about your local soil conditions and climate in

your particular area, you're much more likely to be successful with your meadow, even if it may not look exactly how it did in your mind's eye.

So you've got your local advice, you've got your patch of lawn and you've got the desire to help the wildlife. But how do you actually go about creating a wildflower meadow?

1. STOP FEEDING THE LAWN as the first step. Immediately. Even if you aren't going to create your meadow this year, let the grass rest and go hungry for a year or so. This will make it easier to thin when the time comes and will reduce the nitrogen content of the soil, making wildflowers more likely to flourish. Planting a crop such as yellow rattle in among your grass will weaken it and reduce the nutrient content, giving wildflowers a fighting chance once they are planted.

2. GIVE THE LAWN A GOOD RAKE once it is depleted of some of its vigour and starts to look less healthy, to remove any build-up of thatch at the base. This breaks up the soil surface and exposes bare patches where wildflower seeds can germinate. The more open ground at this stage the better. If you prefer to scrape up the grass then this will work as well, if not better. Grass will inevitably creep back in over time but you'll have a blank canvas for wildflower seed to get a good footing first.

3. SCATTER YOUR RECOMMENDED WILDFLOWER SEED on this roughed up soil in the spring or autumn, either between the grass or on the bare surface. Set aside a few seeds and sow them into trays to create little plug plants. This way if your seeds don't come up you can fill the gaps later on in the season.

4. WATER WELL. Seeds need water to germinate so while they are sitting on the earth make sure they never dry out. Give the patch a water immediately after sowing, unless it rains, and then watch the weather. In spring or autumn there should be plenty of rain but until the first leaves appear on the plants, make sure they are always nice and moist.

5. MONITOR THE MEADOW for the first year. Watch closely to see what species grow well and what species visit. With your spare plugs, fill any gaps that appear and keep in check any plants that look likely to take over. Be aware, however, that each year the dominant species of plants will naturally vary.

6. MOW THE GRASS once all your wildflowers have gone to seed and those seeds have matured. If you cut it before the seeds have matured, you'll find a lot fewer will germinate the following year. This will probably be between the height of summer and autumn. In the summer, mowing will make great hay – early summer to favour next year's spring flowers and late to favour high summer species the following year. A mow in the autumn will weaken the grasses, making room for flowers. You could also mow in the early spring, as this will also knock back very strong grass.

UNSUNG HERO:
BLACK MEDIC

Because of its common name, black medic (*Medicago lupulina*) has some negative connotations. In fact, it is edible and has some medicinal qualities so is very useful. If you struggle with black medic growing in the lawn it can be a pain but it's a sign that the ground is dry and nutrient poor, as opposed to moss growth, which would more likely indicate damp conditions. The good news if black medic shows up, is that you have a very good chance of being able to establish a wildflower meadow as it grows in much the same conditions as many popular species like poppy and cornflower. In fact, black medic itself is a useful plant for a wildflower meadow, as the blooms are highly attractive to bees. As a result it's a large contributing ingredient in lots of honey. The great thing about all plants in this family, which also includes the closely related vetch, clovers and trefoils, is that they fix nitrogen in the soil, so although they grow where there is little nutrient, just their presence increases nutrient levels, allowing for a succession of different wild flowers over time.

· ·

GARDEN FAVOURITE: POPPY

When we think of wildflower meadows the poppy (*Papaver rhoeas*) is the plant that springs to mind. Its cheerful colour, translucent petals and beautiful seed head simply can't be beaten. It is also a poignant flower with its strong association with the First World War. Their abundance during that tragic period was the result of the extreme disruption of the ground during warfare, which brought dormant poppy seeds to the surface where warmth from the sun meant they were able to germinate. This gives you a clue as to the sorts of conditions this plant enjoys. A poor and uneven soil with little competition from other species will be

optimal for their successful germination. This is obviously challenging in a lawn environment, but eliminating more grass will enable the poppy to grow and it's well worth it for the benefits it has to the wildlife. Poppy seeds provide food for a wide variety of species. Do remember that poppies are annual, so growing them one year does not guarantee success thereafter. For best results, let the seeds fully mature – you'll know this has happened when the seed heads turn brown – harvest a few and store them in a dry, cool place. Then re-sow them either directly onto the lawn or into cells and seed trays in the spring.

....

TREES

Trees offer myriad benefits to wildlife. A place to live in safety, high above the vast majority of predators; a place to eat with flowers and fruits at the end of every branch; a place to perch, and a place to shelter from wind and rain. It's not just birds and mammals that shelter in the trees – enormous numbers of insects live on the branches or inside the wood, fungi live in the dead wood, living wood and in symbiosis around the roots, and with algae to form lichen in the branches, where, along with moss, it in turn offers a home for many small invertebrates. Even epiphytic or parasitic species can rely on tree species for their habitat, for example, mistletoe and many types of orchids. So a tree can offer much and with relatively little effort or expense on the part of the gardener. We just need to make sure we choose our trees carefully.

The trees that tend to house the most wildlife are the native species (see page 48). In every country these species will vary slightly, but broadly speaking they will be the species that occur in established, natural woodlands (not forestry land). Although birds can nest in most species, there are a whole host of other life forms living under the surface of the bark in native species. In simple terms, birds and mammals probably won't make their homes in trees where there are no insects, grubs and fungi to feed on. Why would they live in a tree where they can't eat? They would risk life and limb moving from their nest or den to the ground, every time they wanted food. It also means that when they are

eating, they have a long flight or run back to safety, should any predators make an appearance. It simply isn't sensible. So if you want to encourage wildlife, try to choose species that naturally occur in your area.

Similarly, the more food you can get on one species (so to speak) the better. So, when choosing a tree, think about the amount of flowers, fruits and nuts it produces. A tree that flowers can produce a feast of nectar for birds, bees, butterflies and flies alike. Where a herbaceous plant might produce a dozen or so flowers, a tree can produce thousands. If those flowers turn to fruit with a seed or seeds, then so much the better. This will give sustenance to a huge range of different species: wasps, flies, mammals and birds. People don't tend to think of fruit for birds, usually we grow it for ourselves, but crab apples and even apples we eat can provide a nutritious diet that birds love. Flowering trees, otherwise known as broadleaf trees, belong to a group called angiosperms. This group provides a multitude of food for wildlife and some are evergreen. Conifers are gymnosperms and largely evolved before broadleaf trees, so therefore do not produce true flowers, although they usually provide evergreen cover, which offers protection in the winter. Generally speaking, broadleaf species tend to offer more for wildlife than conifers simply because they produce flower and fruit. So if you're limited on space it may be worth bearing this in mind. If you've got plenty of room then go for both conifer and broadleaf species!

BEST WILDLIFE TREES TO INCLUDE IN YOUR GARDEN

TYPE OF TREE	BENEFITS
Cherry, almond, apricot, peach (*Prunus spp*)	blossom for pollinators and fruit for birds
Cherry laurel (*Prunus laurocerasus*)	scented flowers for pollinators and berries for birds
Birch (*Betula spp*)	habitat for more than 300 insects
Apples and crab apples (*Malus*)	blossom for pollinators and fruit for birds
Snowy mespilus (*Amelanchier lamarckii*)	blossom for pollinators and fruit for birds
Hawthorn (*Crataegus*)	late fruit into the autumn and winter
Oak (of all kinds) (*Quercus spp*)	home to hundreds of insects and nuts for mammals in autumn
Fig (*Ficus*)	fruit for birds
Rowan (*Sorbus aucuparia*)	masses of flowers and fruits for pollinators and birds
Whitebeam (*Sorbus aria*)	blossom for pollinators, fruit for birds and leaves eaten by caterpillars
American persimmon (*Diospyros virginiana*)	fruit for birds in warmer climates

Mulberry (*Morus rubra*)	fruits for large and small mammals and birds
Linden (*Tilia x europaea/cordata*)	bring in aphid predators and provide food for aphids, keeping them off your vegetables
Beech (*Fagus sylvatica*)	nuts for small mammals and birds, and great for deadwood species like fungi, lichen and moss
Walnut (*Juglans*)	leaves eaten by caterpillars and moths, and nuts for mammals and birds
Hazel (*Corylus*)	leaves eaten by caterpillars and moths, and coppiced areas are great for butterflies
Sweet chestnut (*Castanea sativa*)	flowers good for insects and nuts eaten by birds and mammals
Tree heather (*Erica arborea*)	abundance of flowers for bees and pollinating insects
Horse chestnut (*Aesculus hippocastanum*)	huge flowers for bees and pollinating insects
Yew (*Taxus baccata*)	provides shelter and hedging for nesting birds and fruit for mammals and birds CAUTION: poisonous, if eaten, to humans, cats and dogs, so avoid if a risk
Pine (*Pinus*)	evergreen shelter for nesting sites, even for large birds of prey like osprey and good for lichen growth
Spruce (*Picea*)	habitat for myriad insects and foliage provides food for some moth species

A WORD ON PETS

We humans love pets, which can be difficult to marry with encouraging wildlife, but it's something that we need to be aware of. Habitat loss the world over has contributed greatly to the struggle of many birds and mammals but the impact of pets cannot be underestimated.

I absolutely love cats, but there is no denying the fact that these wonderful creatures are svelte, athletic and merciless killers. In Australia, for instance, they are considered responsible for the extinction of 33 endemic or native species and it has been argued that their role has been pivotal in as many as 90 per cent of extinctions in the UK (although this is difficult to prove). So, you should consider carefully before breeding or buying cats, and accept that wildlife regeneration will struggle wherever they are. However, there are things you can do to minimize the risks for wildlife, whether it's keeping cats contained in a particular area of the garden, giving them a jangling collar that will warn their prey of their arrival or making sure feeding tables or baths are well out of their reach.

Likewise, dogs can be very problematic for wildlife in the garden. Even though they are less prone to active hunting than cats, just the smell or sound of a dog can stop wildlife from settling comfortably. Unlike cats though, dogs are easily contained, so fencing off your rewilded area will stop them from making a nuisance of themselves and disturbing nests. Another problem with dogs is that we walk them. If you are aware of the wildlife in your area, then you will know hotspots where there are populations of vulnerable species living. In these areas it's wise to keep dogs on a lead or walking by your side, just to stop them from unsettling or disrupting the mating, brooding or nesting of any mammals or birds.

UNSUNG HERO: FUNGUS

Toadstools, mushrooms, mould: these are all words we use for fungi. I am fascinated by fungi and the more I read about their functions in the natural world, the more my fascination grows. The connotations of the word fungus can be negative because of its association with decay, disease and because mushrooms can contain some powerful and highly toxic compounds. But, although they undoubtedly require caution, if we understand these organisms, we begin to respect them as entities.

Fungus is a whole kingdom. Not plant nor animal but something else that is able to interact with both. Unlike plants, which make their food from the sun through photosynthesis, fungus is more like an animal; it needs to feed on other material.

Fungi are often huge organisms. The mushrooms we see are the fruiting bodies where, instead of seeds, spores are produced. The rest, like an iceberg, goes on below the surface and can stretch for miles beneath the soil. It's in the root-like structures called mycelium that our interest lie as botanists. These are made up of a web of tiny threads called hyphae, which attach to the roots of plants to create a symbiotic relationship. The fungus uses starches and nutrients from the plant but, in turn, research suggests that the mycelium provides a form of communication between plants in the same community – warning of an infection, carrying food from plant to plant, enabling a plant to take up nutrients around it and even, it has been claimed, stepping in to help plants fight disease. It is believed that at least 70 per cent of plants rely on these fungal (or mycorrhizal) relationships for survival. So, when you see a mushroom growing, there lies a magical and still largely undiscovered world!

GARDEN FAVOURITE: TITS

From the Paridae family, this group includes great tits, blue tits, crested tits and coal tits among others – some of our favourite garden birds. They have very small beaks and distinctive colouring. With a few exceptions, tits are common garden visitors and have a varied diet, including insects and seeds. They also eat flower buds on occasion when food is scarce. Putting meal worms and seeds out for them should keep them busy and stop them from eating too many of your prized flowers and shrubs. They are regular users of nesting boxes so adding these to trees, fences or buildings will encourage new residents. They are some of the most entertaining birds to observe in the garden as they are very sociable. Watching and listening to them chattering away to each other is a real joy.

HOW TO CARE FOR YOUR TREES

In terms of maintenance, trees are some of the easiest species to look after. If you have done your research about the kind of tree you want to plant for wildlife, then you should also be aware of how large your particular choice grows. Choose carefully and you will almost never need to prune your trees. Perhaps the exception to this is fruit trees, especially if you're growing them against a wall in a fan or espalier (which is still beneficial for wildlife). However, you can buy fruit trees from a specialist grower and can then choose what rootstock you have them on as nearly all fruit trees are grafted. This is where the scion (the top half of the plant) is stuck to a rootstock (the bottom half) and fused together, so you have the same kind of fruit but a slightly different growth habit. This means that you will never have to worry about trees getting too big as they'll be limited by their rootstock; so in these cases pruning will only be for shaping and removing dead or diseased material.

The key consideration is how to plant for success. A tree takes very little looking after once it's established, but ensuring it establishes without a hiccup is very important. There are a few easy rules to follow and you're pretty much guaranteed to have happy, healthy trees.

GOLDEN RULES FOR PLANTING TREES

Nothing could be more functional or more beautiful than a humble tree. Follow these rules, choose your species carefully and you should find that you not only entice wildlife into your garden, but also give them somewhere to stay, to eat, to shelter from wind, rain and hot sun, to hide from predators and to live happily ever after.

1. PLANT TREES IN THE WINTER OR AUTUMN to minimize the risk of shock. If a tree is trying to flower, or produce leaves and grow, the upheaval of being uprooted into a new place with unfamiliar surroundings can be fatal if the conditions aren't favourable. Planting in the dormant season gives the tree a chance to gently wake up in the spring and slowly begin to put out feelers in the roots, anchor itself and find nutrients. Winter planting also means more likelihood of regular rain to water the tree in.

2. USE A MYCORRHIZAL FUNGUS to aid nutrient uptake. It's inexpensive and can be sprinkled onto the roots in a powder form as you plant. These fungal roots attach themselves to the roots of the tree and open the pores, allowing nutrients to be easily assimilated by the tree. It is through fungi that whole woodlands can communicate with each other. We are just discovering the apparent sentience of trees and it is believed that a lot of their ability to 'make decisions' is enabled by fungi (see Unsung Hero: Fungus, page 99).

3. BUY SMALLER SPECIMENS. It is far easier to establish a smaller tree than a large one. It's also much cheaper. You can spend thousands of pounds on a large specimen tree, but do it at your own risk. If you uproot a mature tree that's been happily sitting in a field for five or ten years, it may go into shock and either die or sit and sulk for years before beginning to grow again. A smaller tree will grow more quickly and establish in its new surroundings with relative ease. After five years you usually won't be able to tell the difference between a tree that was planted as a mature specimen and a smaller, more resilient one.

4. DON'T PLANT TOO DEEPLY. A tree that has been buried too deeply will really struggle to establish. With any plant, but particularly with trees, it is best to plant them so that the soil sits at the same level as it did in its flower pot, or even ever so slightly lower. Imagine a tree growing happily in the woods, and you can picture that it usually has a run of roots along the surface of the soil. That's no accident. It sits there because that is where it likes to be. Plants always grow best if we try to mimic their natural habitats.

5. PLANT IT IN A BIG HOLE. A tree that's hugged too closely by compacted soil will struggle to get its roots into its new environment. Nicely broken up soil around the roots will allow new roots to form easily and gain strength before they hit the unbroken ground around the hole.

6. DON'T ADD TOO MUCH COMPOST. A little compost added to the hole will help the tree to establish well. But use too much and you'll find the roots stay only where the compost is and never venture out to find their own nutrients. This means they rely on you to provide water and food, sometimes for years, and their smaller roots system makes them much more likely to topple in high wind.

7. WATER WELL UNTIL THE TREE HAS ESTABLISHED even if it's raining! In order for any plant to settle in, its roots need to make good contact with the soil around it. Pouring a full watering can over a tree when you've first planted it will help that contact to be made and allow it to start growing immediately. For at least the first year, give the tree a regular helping hand with a little extra water in dry seasons. Make sure it's not waterlogged or too wet immediately after planting, though, as this is often even more damaging than a period of prolonged drought.

8. STAKE IT INTO THE PREVAILING WIND at no higher than one-third of the tree's height. Hammering in a stake at a diagonal against the trunk, using a rubber tie to attach the stake to the trunk will allow the tree to be firmly anchored while still giving it a chance to rock in the wind. This rocking motion is hugely important in stimulating cambial growth, which allows a tree trunk to thicken so that eventually (after about three to five years) the stake can be removed and the tree can very happily support itself, even in the wind.

....

GARDEN FAVOURITE: SONGTHRUSH

Songthrushes (*Turdus philomelos*) used to be fairly common and when I first started gardening I would hear them in the leaves, smashing shells on the ground. Nowadays they are on the endangered list and are considered very vulnerable, though we still think of them as being fairly common. The reason they are so good for a wildlife garden is also one of the main reasons they are so in decline; that is their love for eating slugs and snails. We consider slugs and snails as our nemesis when we garden but to a thrush, they make the most delicious meal. However, many gardeners use snail pellets to kill mollusks. These pellets poison slugs and snails and that poison stays in their bodies, so that when thrushes and other birds and mammals and even frogs, eat them, they too succumb to the poison. So to help the songthrush, we can stop using snail and slug pellets. Instead, try using nematodes (microscopic worms) who specifically target that species but affect no others, physical barriers like sheep's wool, egg shells, copper and chilli powder, or beer traps as a last resort. In short, avoid indiscriminately poisoning our wildlife.

....

UNSUNG HERO:
TURTLE DOVE

Historically, pigeons have been treated with suspicion and branded with the nefarious label of 'rodents with wings' because of the popular belief that they cause diseases. However, evidence suggests that the risk of catching a disease from pigeons is extremely low and you are far more likely to catch something from a chicken in fact.

The pigeon family's most vulnerable member is the turtle dove. With its gentle purring sound, this unique bird is on the red list of endangered species, having lost over 70 per cent of its members in Europe since 1980 and over 90 per cent of its UK population.

Turtle doves are migratory, travelling huge distances to Europe to breed for around three months in summer, and then spending winters in sub-Saharan Africa. These short months of breeding are critical and it is believed that intensified farming where grain is harvested early is the reason for this species' sudden downturn. Growing grain like millet, barley, wheat, oat and rye in your gardens will definitely help turtle doves, especially if combined with some thick and tall hedging for them to nest in. Water is also crucial for this species. They produce a milk-like substance to feed their young and drinking plenty of water during this period is key to brooding success.

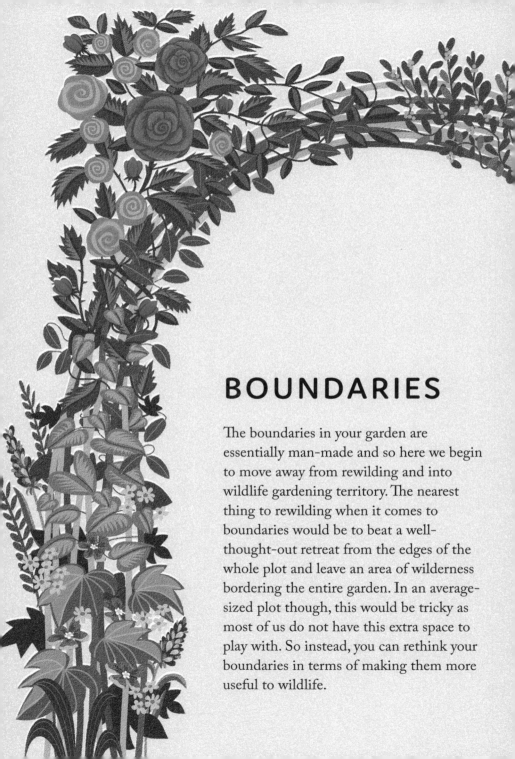

BOUNDARIES

The boundaries in your garden are essentially man-made and so here we begin to move away from rewilding and into wildlife gardening territory. The nearest thing to rewilding when it comes to boundaries would be to beat a well-thought-out retreat from the edges of the whole plot and leave an area of wilderness bordering the entire garden. In an average-sized plot though, this would be tricky as most of us do not have this extra space to play with. So instead, you can rethink your boundaries in terms of making them more useful to wildlife.

WHAT DO I MEAN BY A BOUNDARY?

Essentially, this is anything that marks the end of one area and the beginning of another. It could be the edge of the whole garden or it could be a wall, walkway, fence or hedge that marks the end of one zone, the patio for example, and the beginning of another, like the lawn. Even those of us who have very limited space will invariably have a boundary or two, typically to hide the shed, compost bin or vegetable patch (though I can't think why anyone would hide their kitchen garden!).

If you have very few boundaries, think of adding some. Wildlife actually needs boundaries – in ecological terms, they can act as corridors or pathways that allow animals to move unseen through the space.

As a garden feature, boundaries are incredibly useful. Some of the most famous gardens of the 19th and 20th century, especially during the Arts and Crafts movement, made ingenious use of boundaries. They blocked views with hedges, divided gardens into 'rooms' and gave you little glimpses of what was to come as you moved through the space. Adding boundaries gives you two huge benefits:

1. IN A LARGE SPACE, you can break the area up into bite-sized and manageable chunks that are easier to plan out and plant up, and can allow for different zones and spaces depending on your mood.

2. IN A SMALL GARDEN, they can give the impression of a much larger space hidden around the corner.

GARDEN FAVOURITE: HEDGEHOG

The hedgehog, Europe's native spiny mammal, used to be a common sight but has in recent years ended up on endangered species lists. Widespread use of agricultural pesticides, which kill their sources of food, as well as the removal of habitats like hedgerows and woodland have led to this decline. With increasingly uncertain climatic conditions, we are also finding more confused parents having a late or extra brood, meaning that these hibernating young mammals are often too small to survive the cold winters. If you find a hedgehog in the autumn that weighs less than 500g (and in winter that weighs less than 600g) it should be taken to a rescue centre. You should not disturb a hibernating hedgehog if possible, but always check bonfire piles before you burn them. Hedgehog boxes are great for hibernating hedgehogs in winter and for mothers to raise their young in the spring. If you have one and want to clean it out, you have a short window between the end of winter and the spring to do this. Access is really important for ground-dwelling mammals, so cutting small holes in the base of your garden fences or creating boundaries with hedging instead of solid walls, will enable hedgehogs to move through the neighbourhood – they can cover surprising distances in a single night.

WHAT CAN A BOUNDARY BE?

Practically speaking, a boundary can be absolutely anything, any length, any height, any width and any material. In fact, using as many different materials as possible will maximize the benefits to your local fauna. For example, a low hedge of flowering shrubs could frame the patio, from which you would walk along a path edged by a dry-stone wall, leading into a boundary fence covered with climbers and finally leading to a hedge or bank of small trees at the far end. In this way, while keeping your design varied and interesting, you also allow wildlife to travel in safety from one end of the garden, all the way into next door's garden. You will also provide homes for nesting birds in the hedges, for insects and spiders in the dry-stone wall and food for small mammals, pollinators and birds in the low hedge. Every boundary you have can be enhanced ten-fold by the addition of helpful features to provide shelter and food. Things like insect houses built on walls using twigs, pine cones and stones, espaliered trees or shrubs and climbers, provide refuge for all kinds of species.

Once you've enhanced your existing boundaries, you can add more that bisect these main corridors. Make sure you carefully consider the materials from which they are made. A little row of flowers, grasses or shrubs that lead to the pond or some stacked slate in a curve around the lawn, towards the fence or tree line for instance, will create beautiful features that will enable creatures to access water, food and shelter without having to step a toe in the open landscape. Wildlife will go to great lengths to avoid being exposed in open ground. The more corridors filled with little hidden crevices, bits of moss, ferns, cracks and hidey-holes, the richer and more varied the insect and animal life will be.

With some imagination and creativity, we can create little wildlife havens throughout our garden without even noticing the impact from afar. A well-placed bird box, insect house, trained climber, tree, a few patches of dry-stone wall and hedges as well as fences give our wildlife the essential runs and corridors they need to access different areas of our gardens and also the land immediately outside our plots. It means they can travel from their nests to their food source, unseen by predatory eyes and return to their young with much needed sustenance. It means they can hibernate safely between the cracks and crevices and we need not even know they are there. We don't have to make our boundaries huge, even a small flower border or little hurdle fence or tiny lavender hedge can be the vital thoroughfare that makes all the difference for our garden visitors. We use boundaries to give our gardens structure and form, so it's worth thinking about ways to make them more useful, as well as more beautiful.

UNSUNG HERO: LACEWING

Many people won't have even heard of this beautiful little insect, but keen gardeners know they are a natural predator for mites and aphids. Adults have the delicate, lacy wings, which inspired the name, and they eat some of the smaller insects, but lacewings at the larval stage are super predators. The larvae devour any aphids and mites with ferocious enthusiasm. What is little known though, is that the lacewing adults are also pretty good pollinators. Not bad for such a small insect.

. . . .

UNSUNG HERO: VOLES

These lovely little rodents have long been considered pests by gardeners because of their tendency to chew on seedlings, bulbs and corms and because their burrowing can affect the look of a lawn. Voles are a struggling species. Increased numbers of species like mink and huge pollution of the waterways has made survival for voles (and in particular water voles) very difficult. Having ponds and hedgerows may encourage voles and similar rodents such as shrews into the garden. If trying to attract them is a step too far for you, at least consider no longer seeing them as the enemy. Instead, feel privileged that they have chosen your garden as a place where they feel safe. And maybe put a few wildlife-friendly nets or barriers around prized beds and borders.

. . . .

GARDEN FAVOURITE: FINCHES

A perennial favourite in the garden, the finch family (Fringillidae) is large and varied. These birds come in a huge range of colours with a wide assortment of calls. A few popular finches are the chaffinch, greenfinch, goldfinch, bullfinch, rosefinch, crossbills, redpolls, bramblings, the endangered linnet and hawfinch and the incredibly rare serin. With so many variants, there are a few different things that you can do to encourage them. Most finches eat grubs, insects and seeds but a few can cause a little havoc for your flower buds with their strong and robust bills. Though not all are migratory, the ones that are (such as the Brambling) often visit the UK and Europe in the winter instead of the summer and eat the seeds of trees such as alder, beech and birch, so planting trees like this can encourage visitors. The male and female finches build their nest together and even if they are nesting in a box, they like to add their own lining, so providing plenty of nesting material or growing grasses and fibrous plants can be an added draw.

THE BEAUTY OF HEDGES

Hedges have long been hailed as the key to a wildlife haven and for good reason. They are also fairly low maintenance, if you choose the right species, as one clip a year should keep them looking crisp and neat. Essentially a hedge is a line of plants that is pruned to fit your needs (or not pruned at all, if you pick small species). So in a rewilded scheme, you could let saplings grow and just prune them to the right shape – just be prepared to warn anyone who moves in after you leave that they've got a tamed woodland on their hands!

The best hedges to encourage wildlife involve protection in the form of spikes that prevent larger birds or mammals from easily accessing nests or hiding places. They also have flowers that provide food and preferably go on to produce fruit and seeds that can either be eaten or stored for the winter. If a hedge combines all of these features, then it is going to do a great job and hopefully look good, too. One other key thing to remember when choosing hedgerow plants is that the more variety you have within it, the greater range of wildlife you are going to be able to attract into your garden. Single species hedges give a modern look so if you want a contemporary finish, choose one species for a whole hedge and another species for the next – at least that way you are increasing the range of resources available for the visiting animals and insects. For maximum impact though, a mixed hedge of as many species as you can fit in will be the most effective choice.

BEST SPECIES FOR LARGE HEDGES

Firethorn (*Pyracantha*)	has an abundance of flowers and fruits for insects and birds, with added spikes to make a safe nesting site
Barberry (*Berberis darwinii*)	has spiky branches that provide safe and dense nesting, usually with bright flowers and often colourful foliage
Yew (*Taxus*)	provides evergreen cover for nesting and fruit for birds on female specimens
Holly (*Ilex*)	provides spiky, evergreen nesting safe from predators, and fruits for birds in the winter
Silverberry (*Elaeagnus*)	has evergreen, non-spiky branches for nesting and strongly scented flowers for pollinators
Hawthorn (*Crataegus*)	has an abundance of flowers for pollinators and fruits right through the autumn for birds and mammals
Blackthorn (*Prunus spinosa*)	has an abundance of blossom really early in the season and fruits for birds and mammals in the autumn (sloes)
Viburnum	there are many different kinds of *viburnum*, all with flowers and fruits, some flower in the winter (*bodnantense* and *tinus*) and others in the spring (*opulus*)
Crab apple (*Malus*)	provides blossoms for pollinators and fruits for birds

Dog rose (*Rosa canina*)

BEST SPECIES FOR LARGE HEDGES CONTINUED

Cherry (*Prunus*)	provides blossoms for pollinators and fruits for birds
Cherry laurel (*Prunus laurocerasus*)	gives scented blossoms for pollinators and fruits for birds
Bay (*Laurus nobilis*)	provides a dense, non-spiky evergreen cover for nesting
Dog rose (*Rosa canina*)	has scented blossoms for pollinators and fruits or hips for birds
Cotoneaster	has abundant flowers for pollinators and fruits for birds
Hazel (*Corylus*)	its leaves are eaten by caterpillars and moths, and traditional hedgerows are often laid with branches running horizontally, so it's great for mammal runs
Willow (*Salix*)	this fast-growing plant is great for moths and butterflies, particularly goat willow and pussy willow varieties
Beech (*Fagus*)	not an evergreen but its leaves stay on all winter, so offers safe shelter for nesting birds

BEST SPECIES FOR SMALL HEDGES

Rosemary (*Salvia rosmarinus*)	provides an abundance of flowers for bees
Lavender (*Lavandula*)	provides an abundance of flowers for bees
Currant (*Ribes sanguineum* or *nigrum*)	provides flowers for pollinators and fruits for birds and mammals
Gooseberry (*Ribes uva-crispa*)	provides fruits for birds and spiky branches to provide safe shelter
Ornamental grasses (Poaceae)	gives nesting material for birds and mice (especially evergreen ones)
Small barberry (*Berberis thunbergii*)	has spiky branches for safe movement, shelter and nesting for birds and small mammals
Skimmia	has flowers and fruits (depending on male or female specimens) in winter months
Rose cultivars (*Rosa*)	roses with nectaries and pollen provide food for insects, while roses with hips provide food for birds. Thorny stemmed cultivars can offer safe sheltering spaces

Skimmia (male)

CLIMBERS

If you are after something less expensive and more compact than a hedge, then climbers make a great substitute and can be trained on most upright surfaces along a boundary. They have similar qualities to a hedge in terms of flowering, fruiting and offering physical protection from predators. Some of the best climbers to choose are:

Ivy (*Hedera*)	has late flowers for pollinators in autumn and shelter for birds
Climbing or rambling rose (*Rosa setigera* or *multiflora*, among others)	provides nesting sites, plus potential for pollinators using flowers and birds eating hips, depending on cultivar
Clematis	provides nesting for birds and pollinators can use the flowers
Jasmine (*Jasminum*)	has scented flowers for pollinators
Star jasmine (*Trachelospermum jasminoides*)	has night-scented flowers for night pollinators
Honeysuckle (*Lonicera pericycmenum*)	has strongly scented flowers for all kinds of pollinators, often followed by fruit
Chocolate vine (*Akebia quinata*)	has flowers for pollinators
Grape (*Vitis*)	provides fruit for birds

Ivy (*Hedera*)

Hops (*Humulus lupulus*)

Kiwi (*Actinidia arguta*)	provides fruit for birds and lush shelter and nesting sites – fast growing (you usually need two specimens to have fruit)
Potato vine (*Solanum laxum*)	provides flowers for pollinators
Hops (*Humulus lupulus*)	provides an egg laying site for butterflies
Passion flower (*Passiflora*)	has flowers for pollinators and sweet fruit and seeds for birds

Clematis

HOW TO MAKE A DEAD HEDGE

A dead hedge may not sound appealing, but it is a neat way of displaying your old branches, twigs, cut-back vines and even herbaceous perennials to create corridors for animals to move around in. This method has recently become adopted by gardeners, where previously it tended to only be used in landscape management on a larger scale. And making a dead hedge is really simple. Built up in stages, eventually the whole 'hedge' will be filled with dead wood – great for colonizing fungi, insects and lichen, perfect for mammals to use as a run and for birds to pillage from to build their nests.

1. HAMMER UPRIGHT WOODEN, EXTERIOR TREATED POLES into the ground at regular intervals. The closer together these poles are and the more perpendicular they are, the neater and, therefore, more modern the finished boundary will be.

2. MIRROR THIS LINE OF POLES a minimum of 30cm (12in) from the original posts. You should now have two lines of upright poles running exactly parallel to each other.

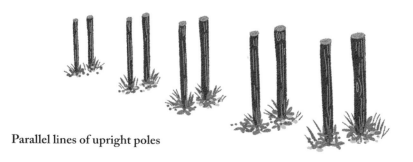

Parallel lines of upright poles

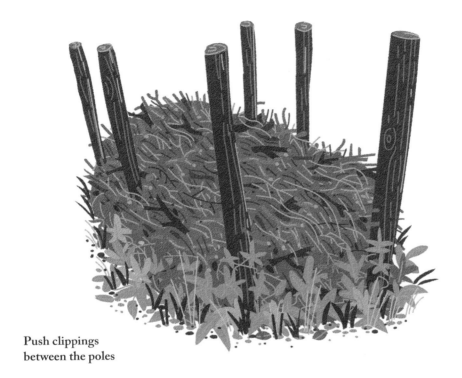

Push clippings
between the poles

3. PUSH CLIPPINGS INTO THE GAP between the two poles every
time you prune a shrub, tree or woody perennial. Note that this can look
really messy unless the clippings are placed in neatly. You can hammer
them down if you want things to be really neat, but remember that the
bigger the gaps inside the hedge, the more creatures will be able to use
the corridor to move around in.

4. ADD A WOODEN TOP connected to the rows of upright poles once
the hedge is full, if you wish to turn this feature into a bench or wall with
a solid top for storing containers or bird baths.

GARDEN FAVOURITE: FOX

For country and city dwellers, the fox has long been considered a pest. But, I love foxes so have to put them in with my favourites. A family of foxes in the garden is a magical thing to witness. Foxes have gained a questionable reputation because of their ability to adapt to almost any surroundings, therefore making themselves very at home in towns and cities as well as in rural areas. They are carnivores and, as such, can be useful in getting rid of things like rats and rabbits in the garden. For me they are magnificent creatures; brazen but elusive. Having a healthy population of all kinds of wildlife in our gardens will invariably bring foxes in to hunt and give them a safe place to stay and even raise their young. Foxes mate in winter and you will hear their distinctive and eerie call (sounding like someone in pain) in the coldest months, if you have them living near you.

. . . .

UNSUNG HERO: GRASSHOPPER

Brought into disrepute by their predilection for destroying cereal crops in huge swarms (when they are known as locusts), on an individual basis, grasshoppers do very little damage, but can still be a slight pest in the garden. However, I've included them in the unsung heroes group (or should that be 'sung' heroes) because of the beautiful ambient sound they make when they rub their legs together. They typically feed on long grass, so growing a patch or two will bring them in where they bathe in the sun and sing their song. Some will feed on other insects, too, so they can be useful. They are naturally shy so normally you won't see them as they will hear you coming and jump out of your way. They have fantastically good ears ... on their bottoms.

. . . .

PATHWAYS

Like boundaries, pathways are an integral part of our gardens in terms of their function but also a useful element to incorporate in terms of design. Pathways can lead us through our outdoor spaces in routes that are surprising and meandering, direct and geometric and even in the form of complex mazes and labyrinths that may have spiritual and historical significance. So, far from being a feature that simply takes you from A to B, when considered carefully and with a little lateral thinking, a path can become an interesting and even exciting feature of the landscape. And like any other feature in the garden, wildlife can be taken into account when considering the shape, routes and materials we make our pathways from. With a few clever omissions or additions, we can create runs and routes for wildlife as well as for ourselves.

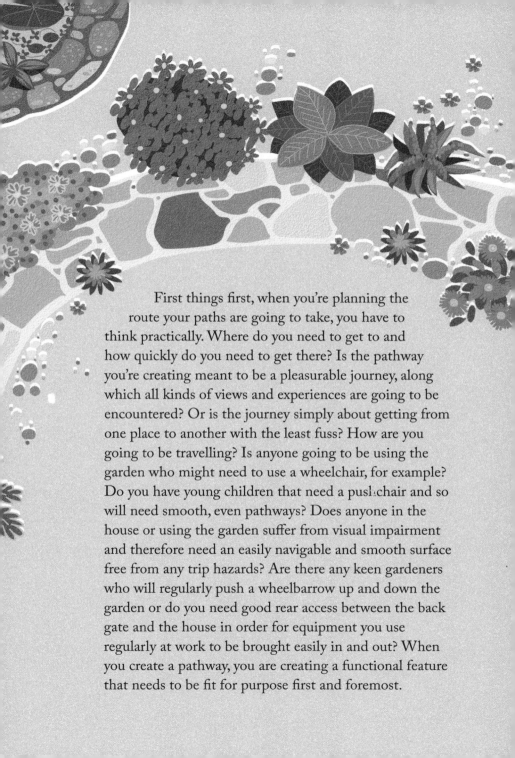

First things first, when you're planning the route your paths are going to take, you have to think practically. Where do you need to get to and how quickly do you need to get there? Is the pathway you're creating meant to be a pleasurable journey, along which all kinds of views and experiences are going to be encountered? Or is the journey simply about getting from one place to another with the least fuss? How are you going to be travelling? Is anyone going to be using the garden who might need to use a wheelchair, for example? Do you have young children that need a pushchair and so will need smooth, even pathways? Does anyone in the house or using the garden suffer from visual impairment and therefore need an easily navigable and smooth surface free from any trip hazards? Are there any keen gardeners who will regularly push a wheelbarrow up and down the garden or do you need good rear access between the back gate and the house in order for equipment you use regularly at work to be brought easily in and out? When you create a pathway, you are creating a functional feature that needs to be fit for purpose first and foremost.

5 TYPES OF PATHWAY TO CONSIDER

Make sure that whatever your pathway is made from and wherever it leads to, it is wide enough to be practical. If it's lawn, for example, ensure it is at least the width of your lawn mower, and whatever the material, it's generally a good idea to make it wide enough for wheelchairs and pushchairs. To make a pathway even more effective for all the creatures who visit your garden, think about combining it with a little hedge, hurdle fence or even a strip of unmown grass, just to give the added bonus of a wildlife corridor.

1. HARD-STANDING

We all know that the more hard-standing and concrete you have in the garden, the more difficult it will be for wildlife, especially subterranean species like worms and other soil organisms. However, if you do need a solid path for any practical reason then of course you must go ahead and build one. In this case, try using permeable membranes beneath the pathway that allow water to soak away and do not hamper the progress of the worms as much. Try to avoid plastic at all costs, but hardcore, rubble, sand, a little cement and some paving blocks or bricks, as long as they don't cover all the surface area of your garden, do not necessarily hinder the wildlife too much.

2. GRAVEL, PEBBLES AND STEPPING STONES

If you don't need a hard flat surface, this is better for soil species. There are many products on the market nowadays that can stabilize the ground beneath loose stones so that wheelchairs, pushchairs and wheelbarrows can easily travel over it, too. Honeycomb structures that sit beneath

UNSUNG HERO: SNAKE

Don't be alarmed, there aren't many snakes our gardens. In fact, if you live in a handful of places including Ireland, Iceland or New Zealand you won't see any at all, as snakes were landlocked after the last ice age thawed. I have only very rarely seen snakes in the garden. Just last year, when visiting a garden I saw a large adder (one of the few snakes that is actually venomous) swimming like a sidewinder through a pond and slithering out the other side. Far from being scared, I was actually awe-struck. It was a magical moment. We so rarely see these elusive creatures and their habitats in the wild are so threatened that I would cherish and protect any snake populations you happen to find in the garden.

gravel will do just this. This kind of pathway will be unproblematic for wildlife, allow water to drain away and will not compromise on the functionality of your pathway or even your driveway if you have one. Warm pebbles will also make a perfect place for lizards and butterflies to bask in the summer sun. Bear in mind that gravel will often fill with weeds, which you can either accept and embrace as wild flowers, or use a less wildlife-friendly membrane (now available plastic-free) underneath the gravel to suppress them. If you want to encourage wildlife then chemicals of any kind are a real no-no, so other methods of weeding will need to be employed, unless you're going for a wild approach of letting them be. Boiling water is a good way of killing weeds without lasting damage. Otherwise, pulling weeds up by hand is one of the simplest options, especially if you keep on top of it and never let them get too much footing.

3. BARK OR WOOD CHIPPINGS

These give a very informal finish but a similar feel and effect to gravel. They are great for woodland-style gardens and in my experience weeds simply don't grow through: the bark or chippings work wonders at suppressing them. Once the chippings break down, you can scrape them up and put them on your flower beds as homemade compost.

4. DECKING OR BOARDWALKS

These are an option that may look sleek and neat but can also provide runs for wildlife. Decking boards are usually laid on top of wooden structures that allow a little gap beneath, where, if you leave a few entry holes, small mammals and insects can run through beneath the surface, completely unseen. Just watch out in wet weather – decking can get very slippery.

5. GRASS

One of the simplest, most beautiful and most wildlife-friendly pathways, is one mown into grass. This can look gorgeous, especially if you are attempting some long grass or a wildflower meadow. In fact, a pathway mown into or around the edge of your long grass can give the whole area a little structure and turn it from a rough and abandoned-looking patch into a beautiful feature in a formal garden, or even in a garden that is being allowed to completely rewild.

UNSUNG HERO:
EARWIG

The earwig is an unnerving insect, with its knack of making us jump as it appears from nooks and crannies, like the inside of a dahlia. However, a little understanding goes a long way and once you know more about the earwig, you might forgive it. They have a tendency to hide as they are in fact nocturnal and sleep during the day in dark places. Earwigs make brilliant mothers, staying with and protecting their eggs all winter, and then nurturing the nymphs until they are independent. Earwigs eat everything, including plants (which can make them a nuisance), but also small insects such as aphids, larvae and crucially, decaying organic matter, so they are incredibly useful in a compost heap, where they will turn your carrot peelings into wonderful mulch in no time.

FORGOTTEN CORNERS

We all have them and sometimes they are not so much forgotten as deliberately ignored, neglected and best not thought about. That little project you've been meaning to do for months, or even years. The compost bin behind the shed, the stack of old pots that's been slowly growing since the dawn of time, that pile of twigs and leaves that you shoved into a corner and then completely forgot about and so on and so forth. They niggle at us with that little nagging voice in our minds every time we think of them. We all need to ignore that voice. Or even better, to tell it it's wrong, because all of those little forgotten places, the piles, the nooks, the never-tackled corners, the I-must-get-round-to jobs, are blissful havens for all the organisms we need but prefer not to think about in our garden ecosystem.

Leaves and logs provide fantastic homes for creepy crawlies of all kinds, but also, and crucially, for fungi that can only survive on dead materials. These areas encourage a huge range of

fungi, which in turn promotes more biodiversity in all species. In the soil, the symbiotic bond between plant and fungus is integral to plant survival. They connect to the roots of plants and interact in ways we are only learning about now. What we do know is that the mycelium (the branched, tubular filaments of fungi) allows plants to access nutrients through their roots. A lot of fungi are delicious to us and other creatures, so there are added benefits to having them.

Compost heaps, neglected or otherwise, make great hideouts and nesting spots for all kinds of animals, insects and mammals. I once found a whole nest of field mice in the compost bin. It's because they are safe from predators, and also very warm. As the micro-organisms in your compost heap break down organic waste they give out a lot of heat energy, so in the cold months all kinds of animals can take refuge in an active compost heap.

The piles of pots and stones and broken bits of furniture around your garden are natural insect homes: a place where they can stay dry and safe and have plenty to feed on. Why destroy that when the creatures living there may be eating the slugs or the aphids who are planning to attach themselves to your most prized specimens?

Obviously you don't want your garden to be overrun with mess and if something really offends you then by all means, remove it. If like me though, you have little forgotten corners that have sat undisturbed for years in a rather tucked-away part of the garden, out of sight and out of mind, then there really is no harm at all in leaving them be. They will provide a vital habitat for all kinds of creatures.

UNSUNG HERO: WOODLICE

These are some of the oldest animal species on the planet, being closely related to the arthropods whose fossils date back about 230 million years. I've personally always loved them. I used to hunt them as a child, under rocks and stones and watch them scurry away or curl into a perfect ball. As I have grown up and learnt about gardening I have found even more respect for this little creature, mainly because of the fantastic work it does in composting. Woodlice feed almost exclusively on decaying plant material so they are some of the most prolific decomposers in the compost heap. They are wonderful little critters.

UNSUNG HERO: SLOW WORM

Although they may look like snakes, slow worms are completely harmless. In fact, they are not snakes, nor worms, but lizards without legs. They often live in dark corners, like compost heaps, and they can consume a large number of plant-eating insects, saving us a lot of work in the garden. They bask in the sunshine, and being cold blooded, can move quickly when warm and slowly if they are cold. They are a protected species so when you find them you must leave them alone and certainly not remove them or disturb their homes. Although there's not an awful lot you can do to encourage them in, the best way of hoping they'll turn up is by leaving little areas undisturbed, especially a warm compost heap, and watch and wait.

. . . .

UNSUNG HERO:
FLIES

We do not tend to think very kindly of flies, but they are a hugely diverse group of insects, from dragonflies to hoverflies to bluebottles. Some we feel warmer towards than others but we should really show respect to the whole range. The main job that flies have in the natural world is not a glamorous one, and that is probably why we feel a little hostility towards them. They are decomposers of decaying bodies, faeces and other detritus. But though this may be a little macabre and repugnant, it is an essential job. It is vital to the survival of the planet and also to us gardeners, for breaking down our compost and turning manure into a useable commodity. Flies of all species (including the common housefly) also help to pollinate many plants, particularly in woodland areas – and unlike bees they are hardy souls undeterred by cold or windy weather.

. . . .

BUILDINGS

We may not all be lucky enough to have space for a building in our gardens but for those of us who do, there are many ways of turning them into wildlife havens in their own right.

An increasingly popular option is the addition of a green roof. Commonly these are planted up with Sedums or other low-growing succulents (you can even buy specially manufactured Sedum matting and turf), but really you can grow anything you like on them, as long as it is shallow rooting and likes the sun. Most roofs get a lot more sun than the ground, simply because they are higher up and less likely to be shaded. Strawberries and lettuces make a great edible roof with some irrigation, or try herbs if you want to save on water. If you want something low maintenance, then a bit of wild grass and wildflower looks absolutely stunning!

HOW TO MAKE A LIVING ROOF

Before you start, make sure that your roof and wall structure is sturdy and strong. A living roof can weigh up to 150kg (330lb) per square metre. Although you can put a green roof on a flat surface, it is best done on a slight pitch of between 2 and 10 degrees. If it's very steep it can be tricky to achieve, so it's worth asking a professional opinion if you're in any doubt. (To make sure you don't fall when making or maintaining a green roof, it's best to have somebody foot the ladder for you whenever you work on it.)

Add a pond liner to make your roof waterproof.

1. Ensure your structure or roof is completely waterproof. Even if you have a watertight roof, add a heavy-weight pond liner, leaving a lip of at least 12cm (5in) around the entire area. You will regret skipping this step if you spring a leak one year in and have to re-lay the whole thing!

2. Build a box-shaped frame the same size as the roof with fairly high sides (minimum 15cm/6in, although deeper rooting plants will need a little extra depth). This can be made out of wood or metal, so pick something you like the look of. If it's

Attach your frame on top of the liner.

wooden, make sure that the wood has been exterior treated. Add some drainage holes of approximately 1.5cm (½in) diameter, spaced roughly 5–10cm (2–4in) apart. It's very important that water can escape into the gutters on the building, so make sure these holes stay clear. Once the roof is planted up, ensure roots never block them by occasionally giving the holes a good poke with a cane. Attach your frame to the roof on top of the liner.

3. Start filling the box with a layer of drainage material. You can put sand at the base to protect the liner if you wish, but gravel should do the trick. Make sure this drainage layer is a couple of centimetres thick at least.

4. Over the gravel, fill the rest of the box with compost or soil. It is best to pick something quite light for

Add a layer of drainage.

this. Soil-based compost tends to be heavy but something like bulrush compost, coir or composted sheep's wool can be added to make things lighter.

Add a light-weight compost.

5. Now you are ready to plant. Depending on the species you have chosen, your planting methods may vary. If you are using Sedum matting and turf, make sure that your compost is nicely tamped down and then roll out strips of planting on top. If you have individual plants, use a trowel to plant exactly as you would in a container.

Start planting up.

6. Water the plants in and make sure they stay well watered until they establish.

7. Maintain your roof once or twice a year by checking that the waterproofing is still effective, the draining holes have not become blocked, weeding where necessary. If you want to, you can cut things back, though this should not really be necessary, especially if you have chosen Sedum.

GARDEN FAVOURITE: BORAGE

As well as being popular with gardeners, this is also a bee's favourite. The borage family has a distinctive acidic, bright blue or ultraviolet colour, sometimes merging into purples and pinks. Viper's bugloss, anchusa, alkanet, and ornamental *Echium* are all members of this regal family. Borage itself is a fantastic plant to grow as a companion for fruit and vegetables. This cheerful annual self seeds so is easy to grow and will bring the bees flocking to your garden. It really is a must for any wildlife gardener.

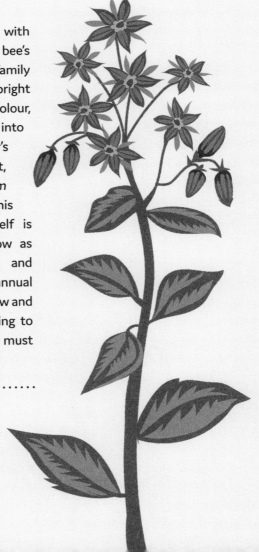

3 WAYS TO MAKE USE OF WALLS

It's not only roof space that's open to some customizing in your garden – the walls on any building are a blank canvas too.

1. PLANTING

If you have brick or stone walls on your outbuildings then planting in them is an effective way of bringing them to life. Avoid structural and load-bearing walls and choose species with shallow roots, like those I've listed here, to prevent any damage to the structure of your walls. Put compost in the cracks between the bricks or stones and plant with *Erigeron karvinskianus* (Mexican fleabane), *Campanula portenschlagiana* (wall bellflower), *Alyssum spinosum* (spring madwort) or *Echeveria* in the sun, or small ferns like *Asplenium trichomanes* (maidenhair spleenwort) or *Asplenium ruta-muraria* (wall rue) and moss in the shade. Plants like this will provide homes and food for insects, mammals and birds as well as making your buildings look rustic and picturesque.

2. ADD SHELVING

If you have a wooden building in the garden then you could very easily and quickly cover the whole of the sides with some shelving. I would recommend creating little boxes of shelves similar to the kinds you find in a wool shop or even a wine shop – that way you can fill each individual nook with something different. Stones in one, straw in another, pine cones next to that, dried wood or bark, twigs, and so on and so on, until the whole thing is stuffed with exactly the kinds of spaces that can be used by a whole host of different insects for nesting and hibernating.

As with any exterior wooden construction, always make sure you use exterior treated wood to build with as it will last a lot longer in the rain and snow. Untreated wood, although great for fungus to develop on, will last only a few years in the elements. I have seen only a few buildings covered like this and really maximizing that space in order to create as many different habitats as possible, but whenever I have come across one I am always struck not only by their ingenuity, but also by their beauty. The plethora of different textures adorned by lace-like, dew-filled cobwebs is mesmerizing to me and for anyone who is serious about providing for wildlife, it's too good an opportunity to miss.

3. HANG BIRD FEEDERS, BATHS OR NESTING BOXES

On virtually any vertical surface you can add some hanging bird feeders on brackets, some hanging bird baths or some bat and bird nesting boxes. You can also buy insect houses that can provide hibernation for solitary bees and other insects. These little sanctuaries, especially if nestled in among some climbers (see Boundaries for a list of the best wildlife climbers, page 120), will immediately attract wildlife and transform a quiet spot into something beautiful and full of activity.

GARDEN FAVOURITE: BAT

We rarely see them but it's a magical thing when we do. Bats eat insects that come out at night and their dung is high in potash, which is incredibly useful to flowering and fruiting plants, so they are very beneficial creatures for gardeners. Housing bats is relatively easy, as wooden bat boxes can be placed on the house, trees or outbuildings if you have any. Bat boxes are best made from untreated, rough wood and placed in a sunny spot that's well out of the way, so that we and our pets will not disturb them. You can encourage nocturnal insects for bats to feed on by planting flowers that are pollinated by moths. Having a pond and compost heap will also enable bats to feed on insects and drink – they often swoop down and drink 'on the wing' from ponds and streams.

FLOWER BEDS

Flower beds are the reason why we all love to garden. There is nothing like walking into an outdoor space, even a tiny courtyard, and being surrounded by the scents and sights of a brimming bed or container full of flowers. And one of the most heady and sensory elements is that when you listen in, a flower bed is nearly always filled with the buzz of bees. Just thinking of that sound, even on the darkest of winter days, fills me with the excitement of a long, warm summer. The obvious conclusion we can draw from it is that pollinators love flower beds as much as we do. The good news is that any flowers will do something to bring in the wildlife. And the more variety the better, so planting lots of different kinds of flowers is the best thing to do.

DESIGNING YOUR FLOWER BEDS
FOR WILDLIFE

Thinking about your flower beds with wildlife in mind goes a little against the grain from what we are taught when we learn to garden and listen to garden designers extolling the virtues of planting in groups of threes and fives. However, the good news is that we can give in to our rebellious instincts to just buy what we like the look of and pack it into the borders. We all do it; we go to the garden centre, see something in full bloom, fall in love, buy it, get it home and think 'oh dear... where am I going to put this?' We feel guilty because we know we should buy and plant in threes and fives, with a limited colour palette but our instincts tell us to buy only what we love and can afford. So right now, when thinking about the birds, bees and beetles, you can rejoice, throw out those 'rules' about taste and design and go with your instincts. Because the wildlife doesn't care what your flower beds look like, they just want plenty of food of all different colours, shapes and sizes, for as much of the year as possible. Planting as many different kinds of flowers as you can fit will provide them with all of this.

UNSUNG HERO: SOLITARY BEE

We all know about honey bees, and that is perhaps because they are so useful to us that we think them vastly important. However, only 9 of the 20,000 species of bee in the world are honey bees and the vast majority are solitary with around 250 bumble bee species alone. Bee-keeping has become such a popular activity that honey bee populations in the UK and Europe are fairly stable. What we do need are solitary bees and these creatures are seriously struggling to find habitats. Some like to nest in trees, others in the ground, some in hollows in walls or between rocks and some in abandoned nests made by other species like rats and mice. So providing varied habitats is key to their survival. The much larger number of solitary bee species means that in fact they do the majority of the pollinating rather than honey bees. Providing varied plant species for these bees to feed from will give them the helping hand they very badly need.

FLOWER BED CONSIDERATIONS FOR WILDLIFE

If you're keen to attract more wildlife into your flower beds, then think about the following factors when you're choosing the flowers you want to plant.

COLOUR is important to us in terms of the design and the mood we want to create in the garden, but to the pollinators aesthetics don't come into it. What matters to them is that they can find the flowers easily. That means that getting a range of colours to enable as many insects to see them as possible is the best way of attracting wildlife. If mixing up all the colours offends your design sensibilities then you can always drift slowly from one range of colours into another as you move through the space. Many insects see the ultraviolet spectrum very clearly, so colours in blues, purples and pinks can often be seen – think of alliums, thistles and lavender – but then many can also see yellows and whites. And

Alliums, thistles and lavender

flowers that are white, especially if scented, can be found by all those night-time pollinators that do their work when we're fast asleep.

FLOWER SHAPE is actually a lot more important than flower colour. Bees, particularly, are attracted to a landing pad-style shape with a single set of petals, often found in flowers in the daisy family. The different colour of the centre from the petals is clearly visible to many pollinating insects. This is not an exclusive rule though and, in fact, some of the most popular flowers for wildlife are in the pea family (things like gorse and broom) and in the nettle family (dead nettles and salvias), which are often teaming with bees, and these have highly unusual shaped flowers known as zygomorphic. Foxgloves are also not typical of the landing pad shape, but the blotches in the throat of the flowers attract and guide the insects in. Add a wide variety of shapes, from tall spires with trumpets like foxgloves or

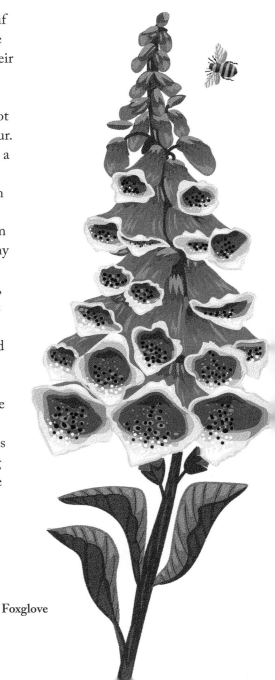

Foxglove

Hawthorn

hollyhocks to densely clustered globes to daisy-like wide open flowers – that way you'll attract the richest diversity of insect.

SCENT is hugely important for guiding insects towards flowers, especially for species that pollinate at night or plants that flower at night, which have evolved to be highly scented and white in colour, so that they can be found in low-light levels. It's a treat for us to have scents wafting through our gardens and fantastically useful for pollinating insects, too. Include flowers like night-scented stocks (*Matthiola longipetala*) or *Nicotiana sylvestris* (ornamental tobacco) and even *Magnolia grandiflora* (scented magnolia), whose scent is strongest at night to attract moths and other nocturnal pollinators.

NECTAR is the reason that insects visit flowers. They will pick up pollen as they go but nectar is the sugary liquid they get from the flowers that will give them sustenance and energy. Not all plants produce nectar though and not all plants produce pollen. So how can you tell if a plant produces nectar and pollen? It's simple, look for the simplest, single flowers. If they are very blousy, petal-filled, double flowers, the anthers (where the pollen is held) and nectaries (where the nectar is held) have been bred out and replaced by extra petals to make them more showy and appealing to humans but rather futile for pollinators. These plants are sterile and do not produce anything useful to wildlife. So the more petals (and the more intensively bred and selected), the less likelihood of pollen.

FLOWERING MONTH is really worth thinking about. In an ideal world you would aim for a flower border to provide food for the wildlife for as much of the year as possible. That means flowers as early in the spring as you can manage and as late into the autumn, hopefully followed by seeds that can sit on the plant throughout the winter. Someone once told me that if you go to the garden centre every month for a year and each time you visit you buy something in flower, you should have flowers in your garden for each season. I think this is a really useful way of shopping for your wildlife borders. If you follow this advice then in theory at least there should be something for the insects all year round.

QUANTITY is always important in flower beds, both in terms of design and practicality. A full, brimming bed will cover the soil, preventing nuisance weeds from taking over, and allow plants to offer each other some natural support instead of you getting out there with stakes and canes. However, plant too densely and some will get out competed, so there is a balance to be struck. Another concern though, is the number of differing species you plant. In short, if you want high numbers of animals and insects you need to give them as much variety as you possibly can. This means planting a mish-mash of many different things. Design dictates that you should limit your species and add multiple and repeated clusters of the same plant. Real life never quite fits into one box or the other, so there really is no problem with combining these two approaches. Add a few of the plants you like in groups and maybe exceed the recommended number of species as dictated by 'good taste' so that there's a little more variety for the visiting creatures.

NATIVE PLANTS often, though not always, offer more to wildlife than plants from far flung corners of the earth. There has been a huge amount of debate and discussion on this subject in recent years but it seems that natives and near-natives (as discussed on page 48) do tend to provide more useful nutrient for pollinators. So anything that comes from roughly the same biome (a climatically similar area) or, in some instances, the same country or even region, will tend to have evolved to offer the same sorts of benefits that your local wildlife needs. Let's not forget that most native wildflowers will be the things we call weeds, so there is a compelling argument for letting some be. Most of us will curate and cultivate some parts of our space and be a little more relaxed in others. Basically, if there is nectar, pollen, seeds and material for birds and mice to make nests with, regardless of where a plant comes from, then chances are it will be used. However, it is definitely worth including some native or near-native plants that have evolved along with your native wildlife and will be most useful to them.

THE NIGHT GARDEN

Some animals only come out under the cover of night – shuffling around in the darkness, we hear their calling and shrieking, which sometimes creates an eerie effect. What we tend not to think about is the implications of those nocturnal activities. Why are they there and what are they doing? The simple answer is, they are doing exactly what other animals do in the day time: feeding and breeding. Some species are active in the evening when scents are at their headiest, and some come to life just before dawn for the opening of single-day flowers such as day lilies (*Hemerocallis*) and morning glories (*Ipomoea*).

When considering nocturnal species, we tend to think of larger animals: owls, foxes, hedgehogs and bats, but these are bio-indicators for the profusion of other, smaller unseen night visitors who we may not notice. These creatures, like moths for example, may not necessarily be the kind of visitors we want, but they play a vital and unique role in the pollination of plants and as such are hugely undervalued.

As the sun sets in warmer countries, you can often see insects like fireflies or hear small frogs and crickets singing and chirping. Even in temperate climes though, night-time activity in the garden is abundant. As wildlife gardeners, our job is not to forget those species, though they may be unseen to us, and to provide them with everything we would for the species we do see; habitat, food and water, safe corridors and nesting sites for breeding and egg laying.

These animals and insects have adapted to thrive in nocturnal conditions when light levels are low, sound carries a long way and scents come to

the fore. Many can see exceptionally well in the dark, but white flowers particularly stand out, so pale flowering plants will often be best for night-time visitors. Other species navigate in the dark using alternative senses where eyes are unreliable. Bats, for example, while they can see, rely on echolocation (a kind of sonar) that allows swift flight around obstacles. For species with poor sight, the night garden is all about scent and there are many plants that have adapted to fill this niche – either with flowers that open at night or scents that suddenly grow in strength in darkness.

Providing a nocturnal ecosystem will cater for moths, nocturnal bees, beetles, grasshoppers, aquatic creatures and flies who will in turn attract bats, mice and hedgehogs, who will bring in the owls and foxes. And so the rich tapestry of life that fills your garden with buzzing and rustling in the daytime, will continue long after we are asleep.

Favourite plants for a night garden include:

- **Night-scented stocks** (*Matthiola longipetala*)
- **Honeysuckle** (*Lonicera periclymenum* or *japonica*)
- **Evening primrose** (*Oenothera biennis*)
- *Magnolia grandiflora* (evergreen magnolia)
- *Trachelospermum jasminoides* (star jasmine)
- *Petunia*
- *Hesperis matronalis* (sweet rocket)
- **Night-scented phlox** (*Zaluzianskya ovata*)
- **Normal phlox** (*Phlox paniculata*)
- *Wisteria*
- *Hydrangea petiolaris* (climbing hydrangea)
- Japanese anemone (*Anemone x hybrida*)
- **Pinks** (*Dianthus*)
- **White-scented daffodils** (*Narcissus thalia*)

GARDEN FAVOURITE: CORNFLOWER

Once a regular sight in fields of corn, hence its name, the cornflower is now very rare in the wild, though we still consider it as being ubiquitous when it comes to wildflower meadows. Cornflowers are annual and need a little coaxing from us (much like poppies), either by roughing up the soil extensively or sowing fresh seed each year and planting out plugs. However, cornflower will readily self-seed if there isn't too much competition. It's well worth making the effort with these plants for their colour, the shape of their buds and flowers and their value for wildlife – a blue that is perpetually visible and appealing to bees in particular.

MAINTAINING YOUR FLOWER BEDS

This really needn't be complicated. As long as you can tell the difference between a shrub and a herbaceous perennial then you'll be fine. Shrubs (and this includes things like lavender and rosemary) have a base of woody stems that it's best to avoid cutting into if you are unsure, and herbaceous perennials tend to be green or at least non-woody right down to the ground, to where they will die back each winter and reshoot from in the spring. These can be cut back hard but, if in doubt, leave 10cm (4in) or so at the base. You may also have grasses in among your flower borders. If so, the deciduous ones can be cut back and the evergreen ones tidied by pulling out the dead blades.

It is really important that when maintaining flowers for wildlife, you leave as much food as possible for creatures through the winter months. Birds, particularly, tend not to hibernate. The non-migratory species that stay through the cold months, or the migratory species that come over for the winter, need food at this time of year just as much as any other and provisions are scarce. So it is really important that you do not cut your flower beds back until the spring. This way the seedheads can sit on the flower stalks for the winter and provide some sustenance for the winter foragers. It is also worth putting out some worms and seed or fruit for birds during the cold months to supplement your flower seeds.

KITCHEN GARDENS

Growing food and providing for wildlife are two things that tend not to go hand in hand. We like to keep our edibles for ourselves and anything else that chooses to nibble on our produce gets quickly eradicated or evicted. So having a kitchen garden or vegetable patch in a space that has been designed to bring in as much wildlife as possible is not necessarily going to leave much for you! Do not despair though. There is plenty that you can do to stop the pests from stealing all of your edibles while still having space for wildlife in the garden.

A very inspiring ecologist I know, called Nadine Mitschunas, taught me a lot about growing produce among an abundance of wildlife, and her lesson was to grow vegetables and fruits hidden in the middle of the beds, with beautiful flowers and ground cover densely planted around the edge. That way birds wandering on the ground will miss the vegetables and only see the flowers and seeds, where they can still get a meal. In turn, this ground cover and flower edge provides homes for predatory insects and beetles that prey on pests like aphids. This kind of planting scheme, which is called companion planting, means that the flowers around the edge attract pollinating insects who go on to pollinate your crops, giving you higher yields and making the plot look gorgeous. Things like *Knautia*, *Phacelia* (a green manure that will feed the beds, too), *Nigella*, marigolds and borage work particularly well. Thyme, woodruff, *Iberis* and poached egg plants (*Limnanthes douglasii*) will provide valuable ground cover.

Having a pond near your vegetables will also make a huge difference to pest number as dragonfly nymphs, tadpoles, frogs and newts will polish off a lot of these, including the dreaded slugs. So having a vegetable patch that is full with all kinds of creatures, means that your crops grow

better and the innate balance of nature keeps things in harmony, preventing any one species from gaining too much dominance and massacring your crops.

There are, of course, some damaging pests that can destroy a whole crop easily – things like rabbits and deer. To avoid these, growing in raised beds and erecting some fences will pay dividends (although fewer frogs will be able to get among your vegetables and protect them from flies and slugs). Growing in raised beds will also protect carrots from carrot fly, which only flies up to about a foot from the ground. Additionally, maintenance is made easier, as is filling your veg-growing space with a nutrient-rich medium like manure or compost.

UNSUNG HERO: NETTLES

Sting they might (although not always), but nettles are incredibly beneficial to wildlife and they are also delicious and full of nutrients for us. Stinging nettles (*Urtica*) can be annual or perennial and there are different sub species, but essentially they all perform the same function and they are all loved by butterflies. In fact, some species of butterfly rely on stinging nettles to lay their eggs. These kinds of nettles, if harvested in early spring, can make a delicious soup or stew. Nettle can also be used in balms, ironically to sooth the skin, or in a bucket full of water to make compost tea (see page 163) and feed other plants. They are truly a super plant. The non-stinging type of nettle, the dead nettle or red dead nettle (*Lamium*) is easily identified by its square stems and little spikes of white or pink flowers and although it's not tasty for us, it produces flowers that are very nutrient rich and valuable to all kinds of pollinating insects.

UNSUNG HERO: TOADS

From fairy tales to modern-day children's films and stories, toads get a very bad rap. This is probably something to do with their appearance; squatter, uglier, wartier and fatter than their cousin the frog. They croak most eerily and they walk cumbersomely rather than hopping elegantly. But the folklore around the toad undoubtedly comes from their poisonous secretions. As far back as the Middle Ages in Europe they are documented as being in league with the devil and have historically been associated with witchcraft both in Europe and further afield, particularly with witch doctors and medicine men in South America and in voodoo rituals. Historically, these poor little animals have been harvested and killed for their secretions, which have been used for both medicinal and recreational purposes. There is a little poison in them, some species more than others, but very little in any European or North American species, and certainly nothing to do much damage. The truth is, toads do a wonderful job at pest control. In a garden setting, their life cycle is very similar to frogs. They live on insects and worms when they are small and as they grow much bigger than frogs, can even feed on small mammals, snakes and slow worms. They have been hammered in number the world over by man's mistreatment but there will always be a welcome home for toads in my garden.

WORKING IN HARMONY
WITH WILDLIFE IN YOUR
KITCHEN GARDEN

We feel the need to protect our crops, but if we are open to all species that we invite to share our gardens then a natural balance should be struck between the pests and their predators. Yes, you might suffer a few losses but not many. Certainly not enough to put you off trying. You might even find things grow better than they did before.

CHEMICALS on your vegetable patch should be avoided at all costs, as this will cause a serious imbalance in the visiting fauna. One pest might be destroyed, allowing another to take its place. Most realistically, in killing the pest you targeted, you will invariably find you kill everything because there simply aren't any effective single-species pest controls out there. Frogs and birds who eat the insects will really struggle to survive. Instead, it is much better to have a self-regulating balance of species to help maintain your garden ecosystem.

DIGGING may be the staple activity of the vegetable gardener, but it is not necessarily the best thing we can do. In fact, every time you dig you disrupt all the complex relationships and processes happening in the soil. It is becoming increasingly encouraged to adopt a no-dig approach in the kitchen garden and the results have been slowly indicating that in fact, doing less work on your soil might actually result in higher yields. So, if you want to do your bit and improve your harvest while maintaining your soil's natural ecosystem, then put the spade down! Have a cup of tea instead...

PERENNIAL VEGETABLES are becoming increasingly popular and easy to get hold of. If you want to stop digging then a method of continually breaking up the soil might be necessary. The roots of perennial vegetables will provide this function. You might be lucky enough to live in a warm climate where you can grow something with hefty roots like cassava, which will break up the soil while maintaining its structure, but for the rest of us, it's artichokes, cardoons, perennial leeks, perennial kale, asparagus, sorrel and many more becoming available every year.

FEEDING with very strong chemicals can be destructive too. Instead, wormeries are a fantastic source of nutrients and can be a really effective way of dealing with our food waste. Simply pour the food in and let the worms do their thing. It will come out fairly quickly as a really rich feed, both a compost-like substance and a liquid, that can be added to your vegetable patches with great efficacy! Natural ingredients are the best things to use as plant food, whether that's ash from the fire, compost, manure that's well rotted or a compost tea. The tea can be made by soaking nettles or comfrey (which both contain high levels of potassium and nitrogen) in water for a couple of months. This can then be diluted in your watering regime without creating a huge problem for the insects and micro-organisms in the soil around your vegetables.

NETTING to protect your fruit trees from hungry birds is an effective way of preventing fruit damage but the wrong choice of netting has enormously negative ramifications for all the local wildlife. Last year on a neighbouring allotment alone, I found a dead mouse, a dead bird, a few trapped or injured birds and a dead snake. These sad discoveries really hammer home that although we might save a few berries from being nibbled, we are taking the life of some local wildlife, who may

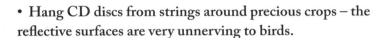

have nesting young to look after. If you absolutely must deter the birds from eating your fruits, then there are alternative preventative methods that are non-destructive (remember though they are going to prevent birds from coming to your garden in general if they work):

• Hang CD discs from strings around precious crops – the reflective surfaces are very unnerving to birds.

• Use a little windmill that spins in the wind as a bird scarer.

• Scarecrows will work for a time, mainly on larger birds. Still, they are fun to make, especially with children, and can certainly work quite well, especially if moved around every now and again!

• If you have to net, then choose the netting carefully: one with holes small enough that you cannot put your finger through, less than 5mm (⅛in). The nets should be white so they can be seen by wildlife at night time, when sadly bats can get fatally tangled in them. Never buy stretchy mesh as small animals can get in or half way in but then the elasticity will cause them to be trapped.

• The kindest method for wildlife is to use fine white mesh bags that can be tied around individual fruits or branches.

HOW TO USE GARLIC FOR SLUG DETERRENT

1. Crush 2 bulbs of garlic
2. Pour on 2 pints of boiling water (3 pints for 3 bulbs etc.) and simmer for 10 minutes.
3. Add water until it's back up to 2 pints and bottle it.
4. Leave to cool.
5. Store in a cool dry place.
6. Every time you spot slug damage, add a couple of tablespoons of the mixture to your watering can and water onto dry leaves in warm weather – though not hot enough to scorch the leaves.
7. After heavy rain you will need to do this again once the leaves have dried as the rain will have washed away the garlic.

TINY SPACES

We are not all blessed with enormous estates. The days of Capability Brown are long gone, and many of us find ourselves renting our houses or our flats without wanting to spend a fortune on a garden that we may only have for a few years. Even if we are able to buy our own houses, gardens have become increasingly tiny or may even be non-existent. You may only have a balcony, tiny courtyard or window sills to work with, but that doesn't mean you can't do your bit for the local wildlife.

4 WAYS TO ATTRACT WILDLIFE
IN A SMALL SPACE

1. PLANTS... even if you only have a window box or two, fill every space that gets some light with containers brimming with nectar-rich flowers. These will benefit pollinators and birds particularly. Plants that are happy in containers such as lavender, thyme, rosemary, lovage, hyssop, broad beans and sunflowers are great sources of pollen and nectar for the bees and really useful in the kitchen too. *Calendula* or *Tagetes* (both marigolds), clary sage and feverfew are great medicinal herbs that also attract a lot of pollinators. Or go fully wild and let the box fill with weeds, some of which – nettles and chickweed – can even become a meal for you if you so desire.

2. BIRD AND BAT BOXES... these make a real difference whether you have a garden or not. All you need is a little bit of wall space and potentially some advice from a local expert or charity as to where to place the boxes to best effect. Providing somewhere safe for some of our most vulnerable species can be a lifeline to them. Remember that the shape and size of boxes will vary according to the species you are looking to home, so always make sure you make or buy one with the correct specifications.

3. INSECT HOUSES... these are best situated outside but they need not be large or extravagant and they are great fun to make with children (see How to Make a Bug Hotel, pages 170–171). Old twigs, bits of stone or slate, pine cones, bricks, holes drilled in bits of wood, leaves, feathers and anything else you can think of will all provide something for one species or another, and you can make it any shape and size. A little box in the corner that acts as a stand for a container or bird bath might be all it takes or you could really go to town and hang something from the wall of the balcony or over the railings. You could turn a bench into an insect house or make a little dolls' house filled with insect-appropriate material. You are only limited by your own imagination – and if you don't have much of an imagination, don't worry, you can let the kids run wild with theirs and take over project managing the insect house construction while you concentrate on the window boxes!

4. FOOD AND WATER... the final thing that we can all do, no matter how big or small our space, from a windowsill, to a balcony, courtyard, roof terrace or even in a huge estate garden, is provide some food and some water throughout the year. Nuts, seeds, fruit, berries, foods with high protein, high fibre and high liquid content are all the best things we can give our birds and mammals and even insects, who often drink from ponds, in order to keep them healthy and able to feed their young. It is vital though, that if you undertake to do this, you keep everything clean to prevent disease and keep provisions topped up regularly so that there's always plenty for the wildlife to find. If you provide food that increases the bird population and that food supply suddenly disappears, the population will be put under a huge amount of pressure. Providing water is also essential for birds and can be done easily in any space. If you have space for a bird bath, make sure the sides are not too steep or high so that any bird or mouse that falls in can climb back out, and also make sure the water is kept clean and changed regularly.

GARDEN FAVOURITE: BLACKBIRD

Where most birds are struggling, the blackbird (*Turdus merula*) is recovering fairly well after a period of decline in the middle of the last century. They remain a favourite species for gardeners, mainly because of their gorgeous song. Urban blackbirds are known for singing in the middle of the night, sadly because they are confused by streetlights into thinking that it's daytime. Contrary to its name, only the male blackbird is black with a bright yellow beak; the female is a dark brown colour. To encourage them into the garden, you need to provide them with plenty of food – and that means worms, insects and fruit in the autumn and winter with berries of bushes like *Cotoneaster*.

HOW TO MAKE A BUG HOTEL

1. Create a structure using wood. This could be made from pallets or crates laid on top of each other or a proper set of shelves. A really easy way to do this is to layer it up as you go.

2. You do not need to put a base on the ground, in fact, if you are building it on bare soil, putting no base will benefit insects the most.

3. Gather materials to fill your structure – feathers, rocks, sticks and twigs, bark, pine cones, bricks and even old bits of broken pots you have lying around are all useful.

4. Pile these materials up bit by bit within your wooden structure.

Using wooden
pallets is a simple
method.

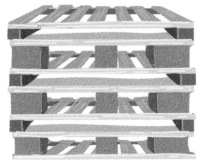

Build up layer by layer.

Fill any holes with
twigs, rocks, broken
pots and bark.

5. For best results group materials together if you can. Just throwing things down will create homes for all kinds of creatures but a hibernating bee will not want to set up home next to a large spider for instance, so keeping similar materials together in distinct areas will mean that everyone will be able to find a safe place. Making large areas for one kind of insect will create more mating potential and increase numbers of creepy crawlies. This ordered approach will also mean your insect house can look more attractive.

6. If you are using the layering method, add a piece of exterior ply wood (with some holes to allow movement) from level to level or an old pallet on top of each level of debris materials. Create another layer of debris and bric-a-brac, and continue up to about 10cm (4in) in height or the height of your pallet.

7. You might want to add something to support the sides of your insect house if you are just laying wood on top of nesting material, but remember that the more access points the insects have, the more creatures you will have using it.

. . . .

GARDEN FAVOURITE: LINSEED

An underestimated, and in many regions, undergrown plant. In southern England where I grew up, down on the coast, and in other parts of Europe, linseed (*Linum usitatissimum*) is grown as a crop by the field by farmers (see all the uses below) and is quite a spectacle. Traditionally linseed is blue but also comes in other colours. Whatever the colour, it is at its most spectacular at dawn when the birds and insects are at their most active. Throughout the day the flowers will slowly fade and die, then new ones will take their place overnight. It is, for me, the sign of summer and it also provides nourishment for the insects at flowering stage and the birds when it's in seed. It's also an incredibly useful plant: the oil is used in art but also as a preservative and to waterproof fabric, which dries solid after linseed has been applied. It is also used to make flax or linen.

. . . .

CONCLUSION

Whether you choose to let the garden go completely wild or make a few positive changes to your approach to gardening, it's worth remembering that even the smallest steps towards considering wildlife can and do make an enormous difference to populations of vulnerable animals, birds, insects, plants and fungi in our outdoor space. With agricultural industries becoming increasingly mechanized and intensified, it's in our gardens that vulnerable species are seeking refuge; here, with varied plants and safe spaces, we are already increasing their numbers. And once nature starts to move back in, and finds a welcome reception and a place to live and eat, you will be surprised how quickly she establishes herself. Start at the bottom, with the grubs, micro-organisms and insects and quickly the frogs, the birds and the mammals will follow. Then provide safe places for them to perform a few of the main stages of their lives; nesting, feeding, egg laying and provide water for them to drink, bath in and live in, ensuring some or all of their life cycle. That may be all you need to do to bring life in and create a busy atmosphere where a rich diversity of creatures and plants can thrive. Then it's just a case of maintaining this without stepping on their toes too much!

MAINTAINING
A WILD GARDEN

Whatever your level of wilderness and however you choose to encourage creatures to thrive in your garden, thinking about how your maintenance regime will impact on the wildlife makes all the difference. After all, it's not just about planning, planting and building, but also the gardening ebbs and flows from season to season. We tweak each year, we plant each year, and each year brings different successes and different failures. It's this daily pottering in the garden that so many of us enjoy. If you love to be busy in your garden then the prospect of letting it all go may not be very appealing. However, even little changes in our behaviour – leaving seed heads a little longer, not eradicating all our weeds – can impact hugely on the success of our rewilded gardens and the creatures that visit them.

SPRING

This is the time for readying the garden for the year to come. Although you should keep disturbance of nesting sites to a minimum, you can take stock of what is and isn't working in your garden. Rethink any spaces that are not doing well and move plants around early in the season. Hopefully, you will have accepted that your wilderness garden will not be quite as neat and pristine as it once may have been, so won't feel the need to pounce on weeds quite so vehemently as in the past. Instead, you can simply enjoy this time in your garden a little more.

• **SOW ANY WILDFLOWER SEEDS** in plug trays to top up your wildflower meadow.

• **DIVIDE MANY HERBACEOUS PERENNIALS** and take soft wood cuttings to propagate extra plants.

• **CUT BACK HERBACEOUS PERENNIALS**. Waiting until the spring allows animals to have maximum use of any seed heads that still remain through the winter.

• **START MOWING PATHWAYS** through any patches of long grass that will form the structure of your garden as the season moves into summer.

• **PLANT CONTAINERS** and fill with nectar-rich flowers, hopefully with scent and bright colours.

• **SOW SEEDS** for the vegetable patch.

• **GROW SUNFLOWERS IN POTS** to either keep that way or plant out into the garden. These seeds are great for the birds.

SUMMER

This is the time for enjoying all of your hard work. The garden and all the creatures visiting it should now be in full swing, with the sound of buzzing and birdsong in the air. Although there are a few tasks still to carry out, you can slow down now and have a chance to relax and enjoy the space.

• **DEADHEAD PLANTS** regularly to keep more flowers coming throughout the season and guarantee there is a rich provision of nectar for pollinating insects. This is particularly true of annual flowers but can also apply to perennials. However, stop deadheading towards the end of summer as seeds on the plants are desperately needed by birds and mammals in the autumn and winter.

• **COLLECT SEED FROM FLOWERS** that you have left. Collect them on a dry day and store them in a cool dry place for the winter, or even in the fridge, ready for next spring.

• **STAKE ANY PLANTS** that are flopping over if you want to. It isn't strictly necessary but if you have let your garden fill with whatever comes its way, things can look a little rough around the edges. Adding a few stakes to prevent the riot of colours and textures from falling all over each other can provide some order.

• **KEEP MOWING** meadow-grass pathways throughout the season and give your meadow its high summer cut.

• **WATER PLANTS REGULARLY** during very dry seasons – especially those in pots. Water in the morning or evening, and do it well and less often rather than lightly and regularly.

• **WEED SENSITIVELY**, remembering that some weeds or wildflowers are really beneficial for the wildlife (see Understanding Weeds, page 40). When you are weeding, do so by hand wherever possible, either using a hoe, scraper, fork or your fingers. Try to avoid using chemicals at all costs as they have such detrimental effects on animal, plant, fungus and insect populations.

• **FEED FLOWERS AND VEGETABLES** regularly with manure, organic feed, homemade compost or compost teas (see page 163). Producing flowers, fruits and seeds is a high energy business and plants need all the help they can get. Having a good soil to start with pays dividends, but even with that, during the crop season, add high-nitrogen fertilizer for leafy vegetables, phosphates for root crops and potash feeds for fruits and flowers.

AUTUMN

Now is the time when the year is beginning to wind down and the garden and all the creatures in it are doing likewise. But far from a time when we can stop, this is exactly when we need to start working. This ensures that come the springtime, when things get in full swing and the animals are busy scurrying around, collecting grains, pollen and laying eggs, we don't run the risk of disturbing them.

• **PILE UP LEAVES** rather than burn or bin them, so that organisms – bacteria, fungi, insects – can inhabit the pile and thrive with plenty of food. Once they have consumed those leaves, what's left makes a great, lightweight and friable compost called leaf mould, which can be used as a mulch for enriching the garden soil.

• **PLACE ANY NESTING BOXES** for hibernating creatures and also for non-hibernating species so that by the time the nesting season comes, the boxes have acclimatized. If they smell new, it can be intimidating or off-putting for ever-cautious wildlife.

• **PLANT ANY BULBS** that will bring early spring nectar sources.

• **LEAVE ANY IVY** until the flowers have well and truly finished. They are some of the last plants to flower and provide a fantastic source of nectar for the last of the season's bees.

• **CREATE HEDGEHOG HOUSES** for any little creatures looking for somewhere to hibernate. Either make a simple wooden box with an open side or door and add some leaves or twigs, or just place some handy piles of leaves and twigs strategically around the garden.

• **ORDER BARE-ROOT PERENNIALS AND TREES**, so that you are prepared for winter planting. Some of these, like hops, are only available in the winter time, so it's best to order in advance.

• **CONTINUE TO FEED BIRDS.** A steady food supply is vital for them in the colder months.

• **CUT YOUR WILDFLOWER MEADOW** once the seeds have all appeared and matured. If you do this before the seeds have formed then you won't have any wildflower the following year. It's not essential, but it is usually helpful to cut at least once in the early autumn or late summer. Cutting long grass can be tricky – most mowers won't manage it – so use a strimmer to get it down to a manageable height.

• **GATHER ANY LONG GRASS CLIPPINGS** and place in a pile somewhere for mice, other mammals and birds to use, especially if they are looking for nest-building materials before hibernation. You can also use these grass clippings to pack in around tender perennials to protect them from the coming frosts or add a little to compost for extra ammonia.

• **RE-SOW ANY PATCHES** in the grass or meadow. Rake up the surface of the bare soil so it's a little more textured, scatter either grass or wildflower seed onto these patches and then water it in. Make sure it stays watered until the seeds begin to germinate.

• **BRING TENDER PLANTS INDOORS** for protection before the frosts begin in earnest.

WINTER

Winter is the time when the rewilding gardener will be busiest, simply because this is the time of year when all the creatures that your garden nurtures will be at their least active. The work you do now will cause them the least upheaval.

• **THINNING PLANTS** is one of the main jobs for this time of year. Even if you are managing a rewilded space, you will invariably find certain plants are taking over and in most cases you will want to keep them slightly at bay. Coppicing and removing any large unwanted species should be done now, when the plants are dormant and the ground has been softened by the rains of the autumn.

• **HEDGE PRUNING** along with general pruning, is best done after the birds have finished nesting. This can be any time from the end of summer right through to the early spring, but generally winter is the best time for this kind of work. Always check the pruning requirements of your hedge species, though, before you start. If you have one that needs a spring prune, try to do this in early spring, after the frost risk is over but before the nesting season is in full swing.

• **WEED** any perennial and offensive weeds that you don't want.

• **CUT BACK** any long grass before insects and frogs have their young, as they will hide in the long grass once they emerge.

• **TACKLE ANY RE-LANDSCAPING PROJECTS** in winter in order to cause minimal disruption to the wildlife in the soil and the plants. Remember that in a wild scheme, the landscaping can add the all-important definition and structure (as well as wildlife friendly opportunities) to your garden.

• **KEEP BIRD FEEDERS** and bird baths well stocked through the tough, colder months.

• **TRY NOT TO DISTURB COMPOST HEAPS** or wood piles in the winter as insects and small mammals will hibernate here. Add compost to the garden in the spring instead.

• **CHECK BONFIRE PILES FOR HEDGEHOGS** before burning.

• **CLEAN OLD FLOWER POTS** ready for use in the spring.

• **REMOVE OLD STAKES** left in flower beds from the summer.

• **REPLANT FLOWER BEDS** in the dormant season.

• **BUY AND PLANT TREES** now while they are cheaper and dormant. Choose bare-root specimens that are only available in the winter and, even if it has rained, water them in and stake them.

• **COVER ANY BEDS FOR VEGETABLES** or hungry flowers with some well-rotted manure as a mulch that will feed the plants through the growing season. Roses, in particular, will thank you for this.

• **REGULARLY CHECK PONDS FOR ICE** and break the surface to allow insects and marine reptiles to breathe down in the depths.

A CALL TO ARMS

It is easy for us gardeners to think of our outdoor spaces only in terms of our own use. However, the very fact that we are starting to think about our gardens' impact on wildlife shows that we are aware of the wider implications our horticultural practices can have. The next step in this is to think of not only our own garden ecosystem but also how that space is a part of the ecosystem at large. Our back yards are just a stop along the road for many species and for wildlife to really recover and thrive, they need many more steps to complete their journeys – a mosaic of different landscapes connected by corridors to provide for all their needs.

It therefore falls on all of us, not only to change the way that we garden, but to try and make an impact on a larger scale, as vulnerable animals, plants and fungi are not able to fight their own battles. Roadsides, railway tracks, hedgerows (which can all act as indispensable corridors enabling the movement of wildlife), farmland, allotment spaces and

other public greenspaces can all play an enormous part in providing stop-offs, nesting and feeding ground for a host of different species.

Once we realize the vast impact we can have just in our own small spaces, I hope we will all begin to challenge those in power and request that they show similar sensitivity to wildlife when maintaining the precious land they are the custodians of. Routine spraying of pesticide and herbicide and regular strimming of these potentially crucial habitats has become par for the course. Instead, we need fewer or no chemicals in use, the acceptance of long grass and wildflower and even the introduction of small wildlife ponds. We can all play our small part in this revolution – which could be the difference between life or death for many a treasured species.

FURTHER READING

Beerling, David, *The Emerald Planet: How Plants Changed Earth's History* (Oxford University Press, 2017)

Goulson, Dave, *The Jungle Garden: or Gardening to Save the Planet* (Vintage, 2020)

Howard, Jules, *The Wildlife Trust, The Wildlife Pond Book* (Bloomsbury Wildlife, 2019)

Kolbert, Elizabeth, *The Sixth Extinction: An Unnatural History* (Bloomsbury, 2015)

Sterry, Paul, *Collins Complete Guide to British Wildlife* (Collins, 2008)

Tree, Isabella, *Wilding: The Return of Nature to a British Farm* (Picador, 2019)

JOURNAL ARTICLES

Burghardt, Karin T., Tallamy, Douglas W. and Shriver, W Gregory, 'Impact of Native Plants on Bird and Butterfly Biodiversity in Suburban Landscapes' in *Conservation Biology* (Volume 23, Issue 1, Wiley-Blackwell, February 2009)

Eastoe How, J., 'The Mycorrhizal Relations of Larch: I. A study of Boletus elegans Schum. in Pure Culture' in *Annals of Botany* (New Series, Volume 4, No. 13, Oxford University Press, January 1940)

Manchester, Sarah J. and Bullock, James M., 'The Impacts of Non-Native Species on UK Biodiversity and the Effectiveness of Control' in *Journal of Applied Ecology* (Volume 37, Issue 5, British Ecological Society, December 2001)

Miyake, Takashi and Yahara, Tetsukazu, 'Theoretical Evaluation of Pollen Transfer by Nocturnal and Diurnal Pollinators: When Should a Flower Open?' in *Oikos* (Volume 86, No. 2, Wiley on behalf of Nordic Society Oikos, August 1999)

Vincent, Delphine, Rafiqi, Maryam and Job, Dominique, 'The Multiple Facets of Plant–Fungal Interactions Revealed Through Plant and Fungal Secretomics' in Frontiers in *Plant Science* (8 January 2020)

OTHER USEFUL RESOURCES

Cornwall Wildlife Trust: cornwallwildlifetrust.org.uk
Freshwater Habitats Trust: freshwaterhabitats.org.uk
Gardeners' World: gardenersworld.com
London Wildlife Trust: wildlondon.org.uk
Merebrook Pond Plants: pondplants.co.uk
Natural History Museum: nhm.ac.uk
Popular Mechanics: popularmechanics.com
RHS: rhs.org.uk
RSPB: rspb.org.uk
The Wildlife Trusts: wildlifetrusts.org
Water Garden Plants: watergardenplants.co.uk
Woodland Trust: woodlandtrust.org.uk
Sciencing: sciencing.com
The Nature Education Knowledge Project: nature.com/scitable/topics

INDEX

ABOUT THE AUTHOR

Frances Tophill is a horticulturist, broadcaster and passionate conservationist. Since 2016 she has been on the presenting team of the BBC's *Gardener's World*. Hailing from Kent she went on to study horticulture at the Royal Botanic Gardens, Edinburgh and has since lived in the south west and south east of England working as a gardener, both in private estates and gardens, and community-based projects.

ACKNOWLEDGEMENTS

I would like to thank everybody at Quercus for offering me an opportunity to write a book that is so close to my heart and has been so inspiring to research and write. To Kerry Enzor and Julia Shone for their guidance through this, a huge thank you. Thank you to Tokiko Morishima and Jo Parry whose design and illustrations have brought the pages to life. Also to Charlotte Robertson, my agent, whose enthusiasm and support is hugely appreciated.

Thank you to those who have inspired my love of nature and respect for all her species; Phil Lusby and Greg Kenicer at RBGE who taught me that there is a wonderful and fragile world of wilderness out there, both on our doorstep and the furthest flung corners. Nadine Mitschunas who opened my eyes to what we can do for wildlife, and Miles Urving, who taught me a huge respect for weeds! To India Hunt whose companionship recently and mutual love for all things ethnobotanical has inspired me – and thank you too, for taking some of my horsetail! Thank you to Fiona Hall for taking me bird watching in all those beautiful places. Finally a big thank you to all those true experts, whose words I have devoured in research and who have opened my eyes to the wonderful world we live in.

DEDICATION: To Mum and Grandma, for walks in the woods and wild education.